Audel™ Electrician's Pocket Manual

Audel™ Electrician's Pocket Manual

All New Second Edition

Paul Rosenberg

Wiley Publishing, Inc.

Vice President and Executive Publisher: Bob Ipsen
Publisher: Joe Wikert
Senior Editor: Katie Feltman
Developmental Editor: Regina Brooks
Editorial Manager: Kathryn A. Malm
Production Editor: Angela Smith
Text Design & Composition: Wiley Composition Services

Published simultaneously in Canada

For general information on our other products and services please contact our Customer Care Department within the United States at (800) 762-2974, outside the United States at (317) 572-3993 or fax (317) 572-4002.

Wiley also publishes its books in a variety of electronic formats. Some content that appears in print may not be available in electronic books.

Library of Congress Cataloging-in-Publication Data: 2003110248

ISBN: 0-764-54199-4

Printed in the United States of America
SKY10055113_091223

Contents

Introduction

In this handbook for electrical installers you will find a great number of directions and suggestions for electrical installations. These should serve to make your work easier, more enjoyable, and better.

But first of all, I want to be sure that every reader of this book is exposed to the primary, essential requirements for electrical installations.

The use of electricity, especially at common line voltages, is inherently dangerous. When used haphazardly, electricity can lead to electrocution or fire. This danger is what led to the development of the National Electrical Code (NEC), and it is what keeps Underwriter's Laboratories in business.

The first real requirement of the NEC is that all work must be done "in a neat and workmanlike manner." This means that the installer must be alert, concerned, and well informed. It is critical that you, as the installer of potentially dangerous equipment, maintain a concern for the people who will be operating the systems you install.

Because of strict regulations, good training, and fairly good enforcement, electrical accidents are fairly rare. But they do happen, and almost anyone who has been in this business for some time can remember deadly fires that began from a wiring flaw.

As the installer, you are responsible for ensuring that the wiring you install in people's homes and workplaces is safe. Be forewarned that the excuse of "I didn't know" will not work for you. If you are not sure that an installation is safe, you have no right to connect it. I am not writing this to scare you, but I do want you to remember that electricity can kill; it must be installed by experts. If you are not willing to expend the necessary effort to ensure the safety of your installations, you should look into another trade—one in which you cannot endanger people's lives.

But the commitment to excellence has its reward. The people in the electrical trade who work like professionals make a steady living and are almost never out of work. They have a lifelong trade and are generally well compensated.

This book is designed to put as much information at your disposal as possible. Where appropriate, we have used italics and other graphic features to help you quickly pick out key phrases and find the sections you are looking for. In addition, we have included a good index that will also help you find things rapidly.

Chapters 1 and 2 of this text cover the basic rules of electricity and electronics. They contain enough detail to help you through almost any difficulty that faces you, short of playing electronic design engineer. They will also serve you well as a review text from time to time.

Chapter 3 explains all common types of electrical drawings, their use and interpretation. This should be very useful on the job site.

Chapters 4 and 5 cover the complex requirements for the installation of motors and generators, and Chapters 6 and 7 will guide you in the transmission of both electrical power and mechanical force.

Chapter 8 covers the very important safety requirements for grounding. The many drawings in this chapter will serve to clarify the requirements for you.

Chapters 9 through 15 cover a variety of topics, such as the installation and operation of contactors and relays, welding methods, transformer installations, circuit wiring, communications wiring, wiring in hazardous locations, and tools and safety.

Following the text of the book, you will find an Appendix containing technical information and conversion factors. These also should be of value to you on the job.

Best wishes,
Paul Rosenberg

1. ELECTRICAL LAWS

An important foundation for all electrical installations is a thorough knowledge of the laws that govern the operation of electricity. The general laws are few and simple, and they will be covered in some depth.

The multiple and various methods of manipulating electrical current with special circuits will not be discussed in this chapter. A number of them will be covered in Chapter 2. Coverage will be restricted to subjects that pertain to wiring for electrical construction and to basic electronics. While there are obviously many other things that can be done with electricity, only those things that pertain to the installers of common electrical systems will be covered.

The Primary Forces

The three primary forces in electricity are voltage, current flow, and resistance. These are the fundamental forces that control every electrical circuit.

Voltage is the force that pushes the current through electrical circuits. The scientific name for voltage is *electromotive force.* It is represented in formulas with the capital letter *E* and is measured in *volts*. The scientific definition of a volt is "the electromotive force necessary to force one ampere of current to flow through a resistance of one ohm."

In comparing electrical systems to water systems, voltage is comparable to water pressure. The more pressure there is, the faster the water will flow through the system. Likewise with electricity, the higher the voltage (electrical pressure), the more current will flow through any electrical system.

Current (which is measured in *amperes,* or amps for short) is the rate of flow of electrical current. The scientific description for current is *intensity of current flow.* It is represented in formulas with the capital letter *I*. The scientific definition of an ampere is a flow of 6.25×10^{23} electrons (called one *coulomb*) per second.

I compares with the rate of flow in a water system, which is typically measured in gallons per minute. In simple terms, electricity is thought to be the flow of electrons through a conductor. Therefore, a circuit that has 9 amps flowing through it will have three times as many electrons flowing through it as does a circuit that has a current of 3 amps.

Resistance is the resistance to the flow of electricity. It is measured in ohms and is represented by the capital of the Greek letter omega (Ω). The plastic covering of a typical electrical conductor has a very high resistance, whereas the copper conductor itself has a very low resistance. The scientific definition of an ohm is "the amount of resistance that will restrict one volt of potential to a current flow of one ampere."

In the example of the water system, you can compare resistance to the use of a very small pipe or a large pipe. If you have a water pressure on your system of 10 lb per square inch, for example, you can expect that a large volume of water would flow through a six-inch-diameter pipe. A much smaller amount of water would flow through a half-inch pipe, however. The half-inch pipe has a much higher resistance to the flow of water than does the six-inch pipe.

Similarly, a circuit with a resistance of 10 ohms (resistance is measured in ohms) would let twice as much current flow as a circuit that has a resistance of 20 ohms. Likewise, a circuit with 4 ohms would allow only half as much current to flow as a circuit with a resistance of 2 ohms.

The term *resistance* is frequently used in a very general sense. Correctly, it is the direct current (dc) component of total resistance. The correct term for total resistance in alternating current (ac) circuits is *impedance.* Like dc resistance, impedance is measured in ohms but is represented by the letter *Z*. Impedance includes not only dc resistance but also *inductive reactance* and *capacitive reactance.* Both inductive reactance and capacitive reactance are also measured in ohms. These will be explained in more detail later in this chapter.

Ohm's Law

From the explanations of the three primary electrical forces, you can see that the three forces have a relationship one to another. (More voltage, more current; less resistance, more current.) These relationships are calculated by using what is called Ohm's Law.

Ohm's Law states the relationships between voltage, current, and resistance. The law explains that in a dc circuit, current is directly proportional to voltage and inversely proportional to resistance. Accordingly, the amount of voltage is equal to the amount of current multiplied by the amount of resistance. Ohm's Law goes on to say that current is equal to voltage divided by resistance and that resistance is equal to voltage divided by current.

These three formulas are shown in Fig. 1-1, along with a diagram to help you remember Ohm's Law. The Ohm's Law circle can easily be used to obtain all three of these formulas.

The method is this: Place your finger over the value that you want to find (E for voltage, I for current, or R for resistance), and the other two values will make up the formula. For example, if you place your finger over the E in the circle, the remainder of the circle will show $I \times R$. If you then multiply the current times the resistance, you will get the value for voltage in the circuit. If you want to find the value for current, you will put your finger over the I in the circle, and then the remainder of the circle will show $E \div R$. So, to find current, you divide voltage by resistance. Last, if you place your finger over the R in the circle, the remaining part of the circle shows $E \div I$. Divide voltage by current to find the value for resistance. These formulas set up by Ohm's Law apply to any electrical circuit, no matter how simple or how complex.

If there is one electrical formula to remember, it is certainly Ohm's Law. The Ohm's Law circle found in Fig. 1-1 makes remembering the formula simple.

Ohm's Law

Voltage = Current × Resistance
Current = Voltage ÷ Resistance
Resistance = Voltage ÷ Current

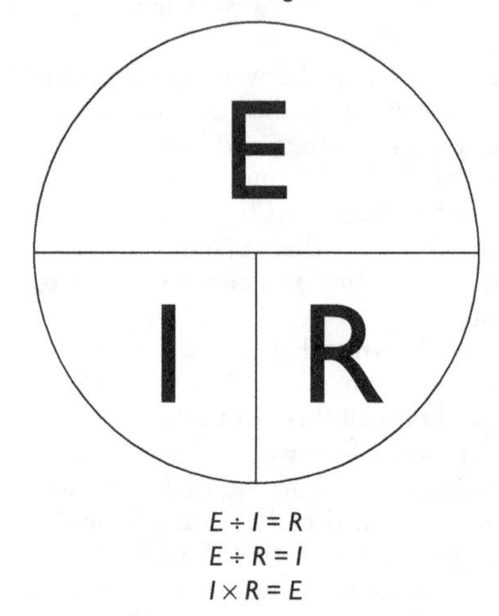

Fig. 1-1 Ohm's Law diagram and formulas.

Watts

Another important electrical term is *watts*. A watt is the unit of electrical power, a measurement of the amount of work performed. For instance, one horsepower equals 746 watts; one kilowatt (the measurement the power companies use on our bills) equals 1000 watts. The most commonly used formula for power (or watts) is voltage times current ($E \times I$).

For example, if a certain circuit has a voltage of 40 volts with 4 amps of current flowing through the circuit, the wattage of that circuit is 160 watts (40×4).

Figure 1-2 shows the Watt's Law circle for figuring power, voltage, and current, similar to the Ohm's Law circle that was used to calculate voltage, current, and resistance. For example, if you know that a certain appliance uses 200 watts and that it operates on 120 volts, you would find the formula $P \div E$ and calculate the current that flows through the appliance, which in this instance comes to 1.67 amps. In all, 12 formulas can be formed by combining Ohm's Law and Watt's Law. These are shown in Fig. 1-3.

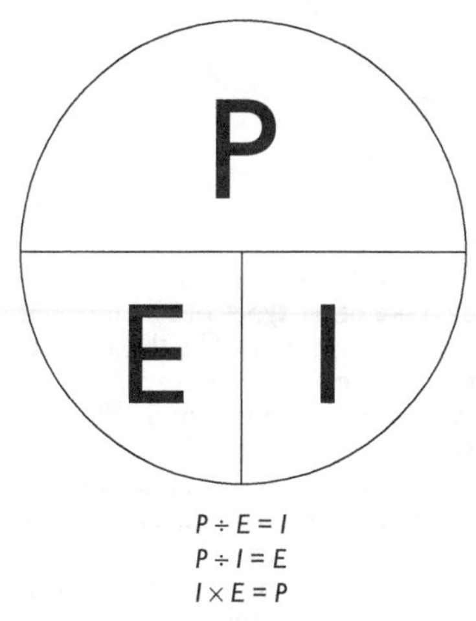

$P \div E = I$
$P \div I = E$
$I \times E = P$

Fig. 1-2 Watt's Law circle.

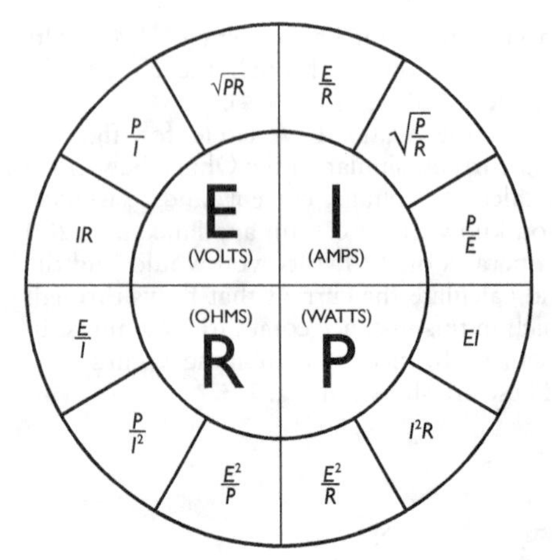

Fig. 1-3 The 12 Watt's Law formulas.

Reactance

Reactance is the part of total resistance that appears in alternating current circuits only. Like other types of resistance, it is measured in ohms. Reactance is represented by the letter X.

There are two types of reactance: inductive reactance and capacitive reactance. Inductive reactance is signified by X_L, and capacitive reactance is signified by X_C.

Inductive reactance (inductance) is the resistance to current flow in an ac circuit due to the effects of inductors in the circuit. Inductors are coils of wire, especially those that are wound on an iron core. Transformers, motors, and fluorescent light ballasts are the most common types of inductors. The effect of inductance is to oppose a change in current in the circuit. Inductance tends to make the current lag behind the voltage in the circuit. In other words, when the voltage begins to rise in the circuit, the current does not begin to rise immediately, but lags behind the voltage a bit. The amount of lag depends on the amount of inductance in the circuit.

The formula for inductive reactance is as follows:

$$X_L = 2\pi FL$$

In this formula, *F* represents the *frequency* (measured in *hertz*) and *L* represents inductance, measured in *henries*. You will notice that according to this formula, the higher the frequency, the greater the inductive reactance. Accordingly, inductive reactance is much more of a problem at high frequencies than at the 60 Hz level.

In many ways, capacitive reactance (capacitance) is the opposite of inductive reactance. It is the resistance to current flow in an ac circuit due to the effects of capacitors in the circuit. The unit for measuring capacitance is the *farad (F)*. Technically, one farad is the amount of capacitance that would allow you to store one *coulomb* (6.25×10^{23}) of electrons under a pressure of one volt. Because the storage of one coulomb under a pressure of one volt is a tremendous amount of capacitance, the capacitors you commonly use are rated in *microfarads* (millionths of a farad).

Capacitance tends to make current lead voltage in a circuit. Note that this is the opposite of inductance, which tends to make current lag. Capacitors are made of two conducting surfaces (generally some type of metal plate or metal foil) that are just slightly separated from each other (see Fig. 1-4). They are not electrically connected. Thus, capacitors can store electrons but cannot allow them to flow from one plate to the other.

In a dc circuit, a capacitor gives almost the same effect as an open circuit. For the first fraction of a second, the capacitor will store electrons, allowing a small current to flow. But after the capacitor is full, no further current can flow because the circuit is incomplete. If the same capacitor is used in an ac circuit, though, it will store electrons for part of the first alternation and then release its electrons and store others when the current reverses direction. Because of this, a capacitor, even though it physically interrupts a circuit, can store enough electrons to keep current moving in the circuit. It acts as a sort of storage buffer in the circuit.

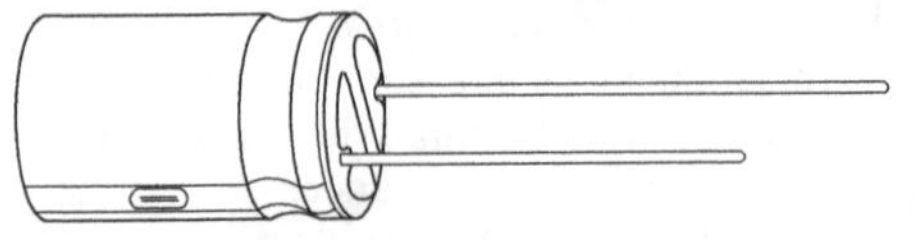

Fig. 1-4 Capacitor.

In the following formula for capacitive reactance, F is frequency and C is capacitance, measured in farads.

$$X_C = \frac{1}{2\pi FC}$$

Impedance

As explained earlier, impedance is very similar to resistance at lower frequencies and is measured in ohms. Impedance is the total resistance in an alternating current circuit. An alternating current circuit contains normal resistance but may also contain certain other types of resistance called *reactance,* which are found only in ac (alternating current) circuits. This reactance comes mainly from the use of magnetic coils (*inductive reactance*) and from the use of capacitors (*capacitive reactance*). The general formula for impedance is as follows:

$$Z = \sqrt{R^2 + (X_L - X_C)^2}$$

This formula applies to all circuits, but specifically to those in which dc resistance, capacitance, and inductance are present.

The general formula for impedance when only dc resistance and inductance are present is this:

$$Z = \sqrt{R^2 + X_L^2}$$

The general formula for impedance when only dc resistance and capacitance are present is this:

$$Z = \sqrt{R^2 + X_C^2}$$

Resonance

Resonance is the condition that occurs when the inductive reactance and capacitive reactance in a circuit are equal. When this happens, the two reactances cancel each other, leaving the circuit with no impedance except for whatever dc resistance exists in the circuit. Thus, very large currents are possible in resonant circuits.

Resonance is commonly used for filter circuits or for tuned circuits. By designing a circuit that will be resonant at a certain frequency, only the current of that frequency will flow freely in the circuit. Currents of all other frequencies will be subjected to much higher impedances and will thus be greatly reduced or essentially eliminated. This is how a radio receiver can tune in one station at a time. The capacitance or inductance is adjusted until the circuit is resonant at the desired frequency. Thus, the desired frequency flows through the circuit and all others are shunned. Parallel resonances occur at the same frequencies and values as do series resonances.

In the following formula for resonances, F_R is the frequency of resonance, L is inductance measured in henries, and C is capacitance measured in farads.

$$F_R = \frac{1}{2\pi\sqrt{LC}}$$

The simplest circuits are series circuits—circuits that have only one path in which current can flow, as shown in Fig. 1-5. Notice that all of the components in this circuit are connected end-to-end in a series.

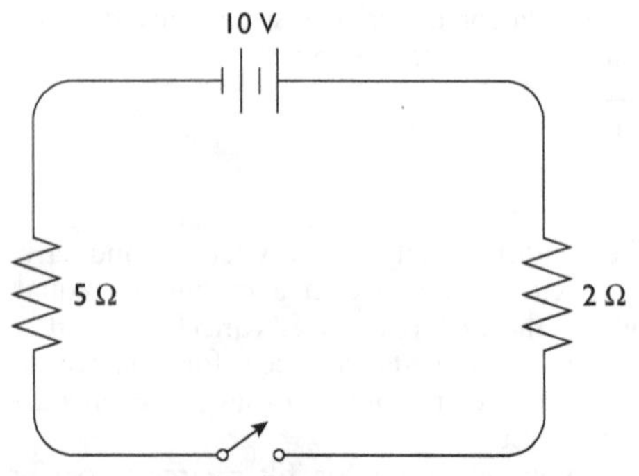

Fig. 1-5 Series circuit.

Series Circuits

Voltage

The most important and basic law of series circuits is *Kirchhoff's Law.* It states that the sum of all voltages in a series circuit equals zero. This means that the voltage of a source will be equal to the total of voltage drops (which are of opposite polarity) in the circuit. In simple and practical terms, the sum of voltage drops in the circuit will always equal the voltage of the source.

Current

The second law for series circuits is really just common sense—that the current is the same in all parts of the circuit. If the circuit has only one path, what flows through one part will flow through all parts.

Resistance

In series circuits, dc resistances are additive, as shown in Fig. 1-6. The formula is this:

$$R_T = R_1 + R_2 + R_3 + R_4 + R_5$$

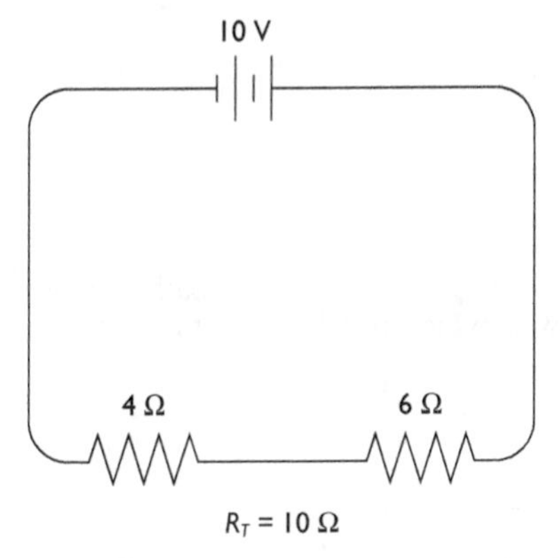

Fig. 1-6 dc resistances in a series circuit.

Capacitive Reactance

To calculate the value of capacitive reactance for capacitors connected in series, use the product-over-sum method (for two capacitances only) or the reciprocal-of-the-reciprocals method (for any number of capacitances). The formula for the product-over-sum method is as follows:

$$X_T = \frac{X_1 \times X_2}{X_1 + X_2}$$

The formula for the reciprocal-of-the-reciprocals method is this:

$$X_T = \frac{1}{\frac{1}{X_1} + \frac{1}{X_2} + \frac{1}{X_3} + \frac{1}{X_4} + \frac{1}{X_5}}$$

Inductive Reactance

In series circuits, inductive reactance is additive. Thus, in a series circuit:

$$X_T = X_1 + X_2 + X_3 + X_4 + X_5$$

Parallel Circuits

A parallel circuit is one that has more than one path through which current will flow. A typical parallel circuit is shown in Fig. 1-7.

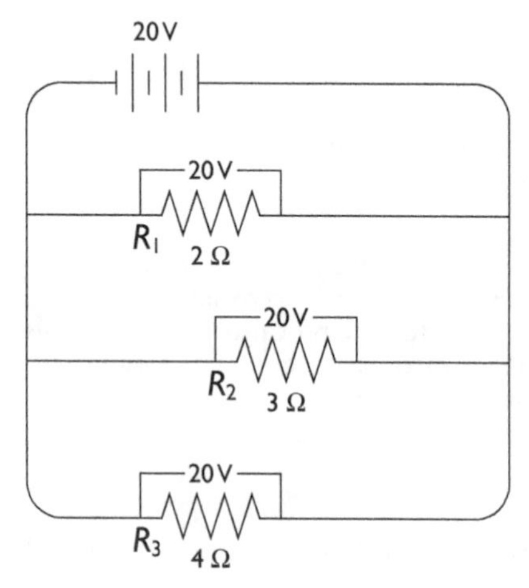

Fig. 1-7 Parallel circuit.

Voltage

In parallel circuits with only one power source (as shown in Fig. 1-7), the voltage is the same in every branch of the circuit.

Current

In parallel circuits, the amperage (level of current flow) in the branches adds to equal the total current level seen by the power source. Fig. 1-8 shows this in diagrammatic form.

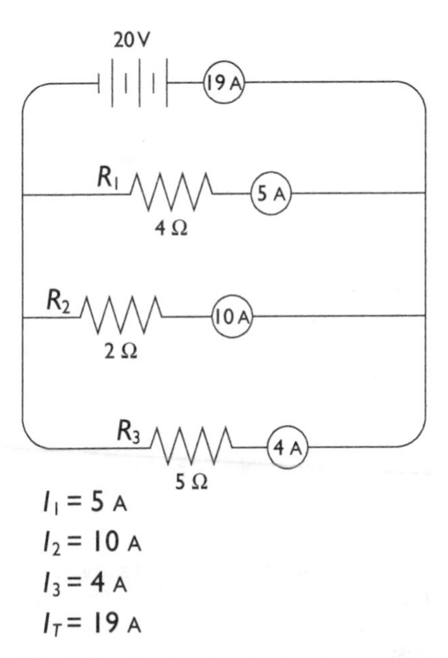

Fig. 1-8 Parallel circuit, showing current values.

Resistance

In parallel circuits, resistance is calculated by either the product-over-sum method (for two resistances):

$$R_T = \frac{R_1 \times R_2}{R_1 + R_2}$$

Or by the reciprocal-of-the-reciprocals method (for any number of resistances):

$$R_T = \frac{1}{\frac{1}{R_1} + \frac{1}{R_2} + \frac{1}{R_3} + \frac{1}{R_4} + \frac{1}{R_5}}$$

Or, if the circuit has only branches with equal resistances:

$$R_T = R_{\text{BRANCH}} \div \text{number of equal branches}$$

The result of these calculations is that the resistance of a parallel circuit is always less than the resistance of any one branch.

Capacitive Reactance

In series circuits, capacitances are additive. For an example, refer to Fig. 1-9. Notice that each branch has a capacitance of 100 microfarad ("mfd" or "μf," written with the Greek letter mu (μ), meaning *micro*). If the circuit has 4 branches, each of 100 mfd, the total capacitance is 400 mfd.

Inductive Reactance

In parallel circuits, inductances are calculated by the product-over-sum or the reciprocal-of-the-reciprocals methods.

Series-Parallel Circuits

Circuits that combine both series and parallel paths are obviously more complex than either series or parallel circuits. In general, the rules for series circuits apply to the parts of these circuits that are in series; the parallel rules apply to the parts of the circuits that are in parallel.

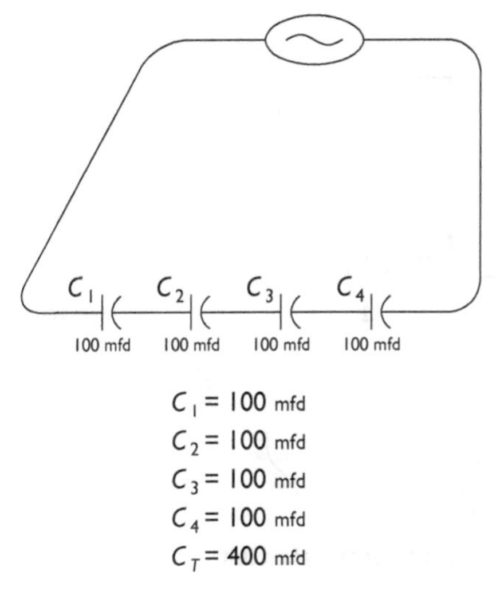

Fig. 1-9 Capacitances in a series circuit.

A few clarifications follow:

Voltage

Although all branches of a parallel circuit are exposed to the same source voltage, the voltage drops in each branch will always equal the source voltage (see Fig. 1-10).

Current

Current is uniform within each series branch, whereas the total of all branches equals the total current of the source.

Resistance

Resistance is additive in the series branches, with the total resistance less than that of any one branch.

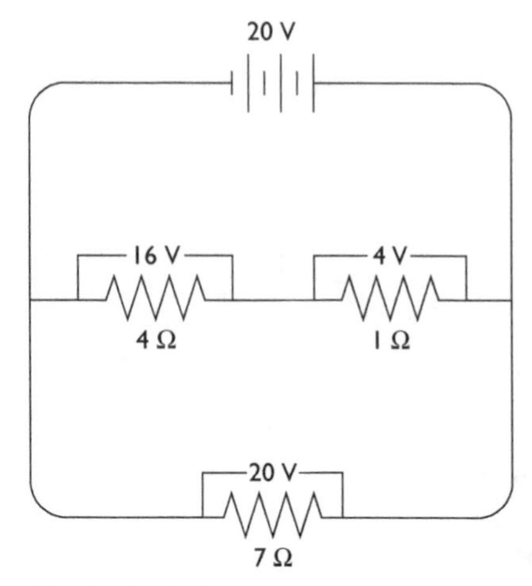

Fig. 1-10 Voltages in a series-parallel circuit.

Capacitive Reactance

X_C is calculated by the reciprocal-of-the-reciprocals method within a series branch, and the total X_C of the branches is additive.

Inductive Reactance

X_L is additive in the series branches, with the total inductive reactance less than that of any one branch.

Power Wiring

Nearly all power wiring is connected in parallel, so that all loads are exposed to the full line voltage. Loads connected in series would experience only part of the line voltage.

One of the most widely used calculations for power installations is simply to calculate amperage when only voltage and power are known. (See Fig. 1-2 and the associated discussion.)

For power wiring, capacitance is rarely a problem. One exception is that long runs of cables can develop a significant level of capacitance either between the conductors or between one or more of the conductors and a metal conduit encasing them. A proper grounding system will normally drain such a charge. If, however, there is a flaw in the grounding system, such as a bonding jumper not properly connected, strange voltages can show up in the system. These voltages are called *phantom voltages*.

Inductance, unlike capacitance, is a serious problem in power wiring. Inductive reactance causes a difficulty with a wiring system's *power factor*. This will be covered in some depth in Chapter 7.

2. ELECTRONIC COMPONENTS AND CIRCUITS

The first thing to remember about electronics is that the laws that govern the operation of electricity (that is, Ohm's Law, Kirchhoff's Law, Watt's Law, calculations of parallel resistance, etc.) are the same laws that govern electronics. In reality, working with electronics is not that different from many types of electrical work. The main differences are the amount of power being used and the exotic-sounding names of electronic components.

To many people, the names of the devices are especially intimidating: Zener diodes, field-effect transistors, PNP junctions, and so on. When you realize that these are little more than fancy names for such things as automatic switches, a lot of the mystery evaporates. Actually, these devices are not especially difficult to understand and use.

Advantages

Electronic circuits possess five basic abilities that normal electrical circuits don't. All of the other amazing abilities that electronic products have are merely combinations of these five.

1. Electronic devices can respond to very small signals and from them can produce a much larger signal. This is how transistors can amplify signals.
2. Electronic devices can respond much, much faster than can electrical devices such as relays.
3. When operating at high speeds, electronic devices can produce magnetic signals such as radio waves, X-rays, or microwaves.
4. Certain types of electronic devices can respond to light. A good example of this is found in common photocells.
5. Electronic devices can control the direction of current flow.

Tubes and Semiconductors

The five abilities just mentioned were first evident in vacuum tubes, long before anyone had heard of semiconductors. Without vacuum tubes, radio, TV, X-rays, and a host of other things would have been impossible. These tubes were the first electronic devices. They took time to heat up before they could operate, they often burnt out, and they were relatively expensive. Nevertheless, they could do things that no electrical device could do, and thus they were very widely used. Even the first computers were composed of vacuum tubes.

Semiconductors, however, have gone a step beyond. First of all, semiconductors do virtually all of the jobs that electron (vacuum) tubes do, plus a few extra jobs—and they operate more efficiently. They don't need to warm up before they can operate, and they are very small. The first computer filled up a space the size of a large garage due to the large size of the tubes. With the small size of semiconductor devices today, you can fit a far, far more powerful computer on a desktop. In the case of the computer, the tubes and semiconductors primarily did the same jobs, but the size difference was extremely important.

One more step was critical: developing the means to put hundreds of semiconductors on one small piece of silicon. This device—the integrated circuit chip—is merely a large number of semiconductor devices squeezed into a very small area. Needless to say, the IC chip has had a major impact on the modern world.

The invention of the electronic tube was crucial to many of the most important developments of the first half of the twentieth century; likewise, semiconductors and IC chips were critical to developments in the last half of the twentieth century.

Semiconductors

In the electrical field, you are familiar with conductors such as copper and aluminum wires and buses. There are also nonconductors (usually called insulators) such as rubber, plastics, and mica. Semiconductors are the materials somewhere in between conductors and nonconductors—that is, *semi*-conductors. In other words, they conduct electricity partially or under certain circumstances.

If you've ever had an electrical theory class, you will remember that an atom can have a maximum of eight electrons in its outer electron shell. You also learned that because electricity is a flow of electrons, atoms with only one electron in their outer shell are good conductors because one lone electron can be shaken loose from an atom fairly easily. You also found out that electrons are very hard to remove from an atom that has seven or eight electrons in its outer shell. Therefore, atoms that have seven or eight electrons in their outer shell are said to be nonconductors.

Semiconductors are atoms that have four electrons in their outer shells. These elements are silicon, germanium, and tin. When one element with three electrons in its outer shell and another element with five electrons in its outer shell are mixed together, they give the resulting compound an average of four outer-shell electrons, making that compound a semiconductor. This is the case with gallium arsenide, a combination of gallium and arsenic.

Silicon and germanium are the two materials that are commonly used as semiconductors. But in their pure form, these materials are not very useful. They conduct a little bit of electricity and not much more. It is when you modify these substances that they become interesting.

You modify silicon and germanium by adding small amounts of other materials to them. This is called *doping*. When properly done, doping gives a semiconductor either a surplus of electrons (making it a type N semiconductor with extra electrons that carry a negative charge) or a deficiency of electrons (making it a type P semiconductor with a positive bias because of the lack of electrons). You may want to take a moment to review this paragraph to grasp all the important details fully.

Now, the idea of a PN junction is simple: It is merely the place where type P and type N semiconductors are placed together. The idea of an NPN semiconductor is also easy: It is merely a sandwich with type N layers on the outside and a type P layer in the middle.

Diodes

A diode is simply a PN junction: a piece of type N semiconductor joined to a piece of type P semiconductor. (See Fig. 2-1.) If you connect a battery to the diode as shown (positive terminal to N, negative terminal to P), no current will flow through the diode with the exception of a very small "leakage" current.

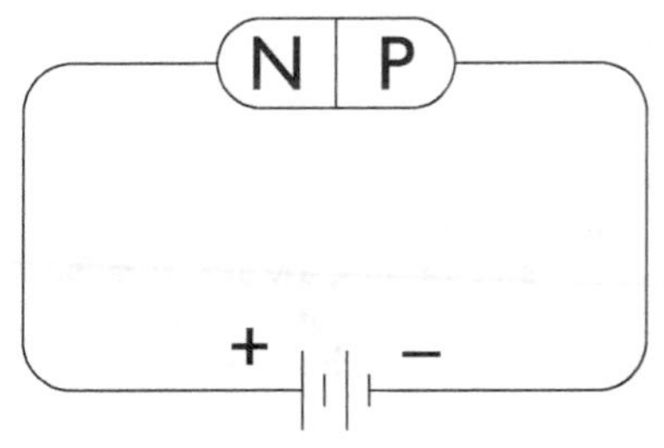

Fig. 2-1 Reverse-biased diode.

Now, if you look at Fig. 2-2, you see the same diode connected the opposite way—with the positive terminal to P and the negative terminal to N. When connected this way, current will flow with very little resistance.

Figure 2-1 shows the diode connected to the battery in a way that makes it *reverse-biased*. This means that it is connected so that it opposes current flow—its bias is reversed.

Fig. 2-2 shows the diode connected to the battery in such a way as to make it *forward-biased*. This connection allows the current to move forward through it.

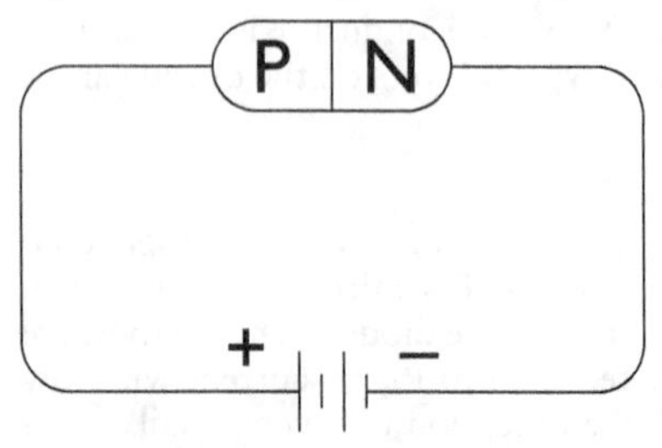

Fig. 2-2 Forward-biased diode.

Diodes are commonly used to convert alternating current into direct current. By simply connecting the diode in series with a circuit, you allow current to flow in only one direction; it won't flow in reverse. Thus, the current can no longer alternate; it can flow in only one direction.

Diodes come in all sizes and ratings. (Make sure you don't connect a diode rated for 24 volts on a 120-volt circuit!) Usually, diodes look like resistors, but they can come in varied sizes and shapes.

What a Transistor Is

After you take away all of the mystique surrounding the "transfer resistor," which is what you now call the transistor, you find that it is an automatic switch. It's a pretty impressive automatic switch, to be sure—but essentially it's just an automatic switch.

The basic transistor is an NPN junction in which one side is more heavily "doped" than the other side. In other words, one of the N sides is more negative than the other N side. The more heavily doped side is called the *emitter*, and the less heavily doped side is called the *collector*. The P section that is sandwiched in between is called the *base*. This is shown in Fig. 2-3.

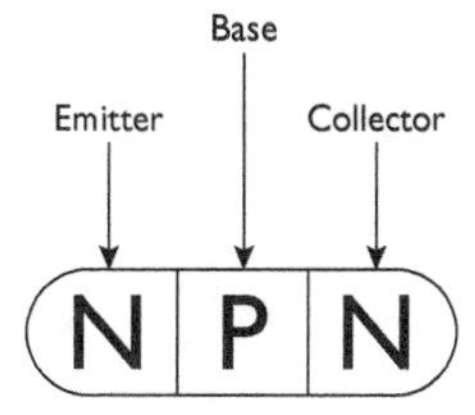

Fig. 2-3 NPN transistor.

To understand how this device works as an automatic switch, look now at Fig. 2-4. As shown in this figure, you will connect the same transistor in a circuit. Looking at the right side of Fig. 2-4, you see that the collector-to-base NP junction is reverse-biased. (Refer to Fig. 2-1 again.) Therefore, except for a very small leakage current, no current flows through this junction. Now, looking at the left half of Fig. 2-4, you see that with the switch open no current flows in that part of the circuit either.

So far, so good. But when you close the switch, something unique happens. As more current flows through the base-to-emitter NP junction (on the left side of Fig. 2-4), it changes the charges in the other NP junction and allows current to flow through it, too. If current flows in the left side of the circuit (base-to-emitter), current will flow through the right side also (collector-to-emitter). If no current flows in the left (base-to-emitter) side, none will flow in the right (collector-to-emitter) side either.

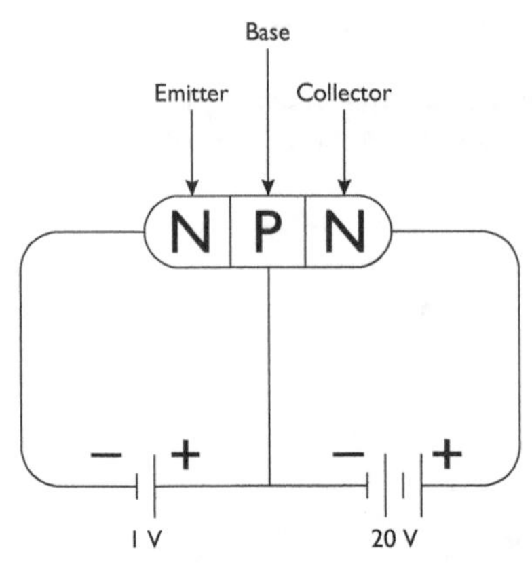

Fig. 2-4 NPN transistor connected in a circuit.

The scientific explanation of why and how the second NP junction changes to allow current to flow is a difficult one. For this book, let it be sufficient to accept the fact that it does work.

If you look at Fig. 2-4 again, you see that the voltage of the battery supplying power on the left side of the diagram is only 1 volt, but the voltage on the right side is 20 volts. So, with this circuit, you can use a 1-volt circuit to control a 20-volt circuit. This is a basic amplifier.

Now, to make it really interesting, here is one last thing that the transistor does: It keeps the current that flows through the right side of our circuit proportional to the current level in the left side of the circuit. In other words, if 5 milliamps flow through the left side, allowing 100 milliamps to flow through the right side, then increasing the current in the left side to 10 milliamps will automatically increase the current in the right side to 200 milliamps. (You are assuming here that all other things remain unchanged.) This relationship is shown in Fig. 2-5.

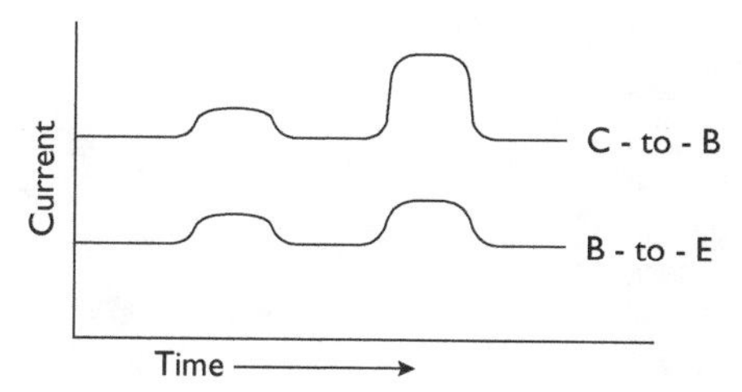

Fig. 2-5 Current relationships in transistor circuit.

You can see from this description how useful transistors are. And considering that they can be produced in extremely small sizes, they become much more important.

Silicon-Controlled Rectifiers

These devices, which are usually called *SCRs*, are composed of four layers of silicon P and N semiconductors (see Fig. 2-6). Unless current is put through the gate lead of the device, no current will flow from the anode to the cathode. If there is a gate current, the resistance between the anode and the cathode drops to almost zero, allowing current to flow freely. Thus, the gate current is necessary to start the rest of the SCR conducting. Unlike the transistor, however, the current will continue to flow from the anode to the cathode, even when the gate current ceases. Once started, the anode-to-cathode current will flow until it stops on its own; it won't be stopped by the SCR.

SCRs are particularly useful because they can handle large amounts of current, especially as compared to other solid-state devices. Commonly available SCRs can handle continuous currents of hundreds of amps.

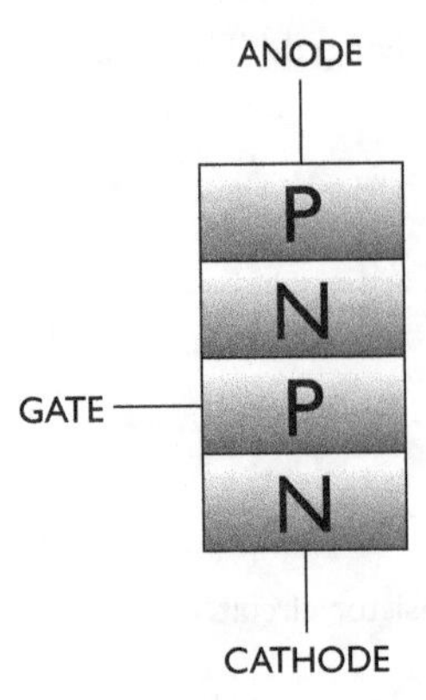

Fig. 2-6 Silicon-controlled rectifier.

Triacs

Triacs are modified SCRs, as shown in Fig. 2-7. The triac blocks current flow in either direction until a current is sent in or out of its gate. Once one of these currents begins, current will be allowed to flow in either direction through the triac.

Triacs are the functional components inside most dimmer switches and similar devices.

Field-Effect Transistors

Field-effect transistors use type P semiconductors on both sides of a type N semiconductor to act as a gate. The P semiconductor gate controls current flowing through the type N semiconductor.

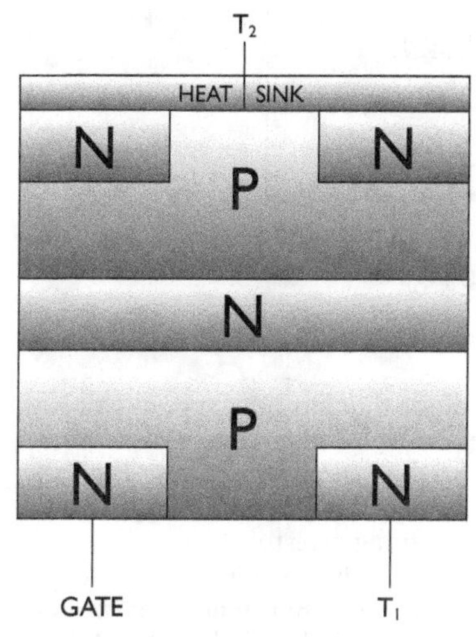

Fig. 2-7 Triac.

Figure 2-8 shows a field-effect transistor. In this transistor, the type N semiconductor will carry the current that you want to control. If you place a voltage on the type P semiconductor (the gate), no current will be allowed to flow through the type N semiconductor. The voltage placed on the P sections creates an electrostatic field that alters the charges in the type N semiconductor, disallowing the passage of current. When the voltage on the P sections is eliminated or reduced, current will be able to pass from the source to the drain of the field-effect transistor.

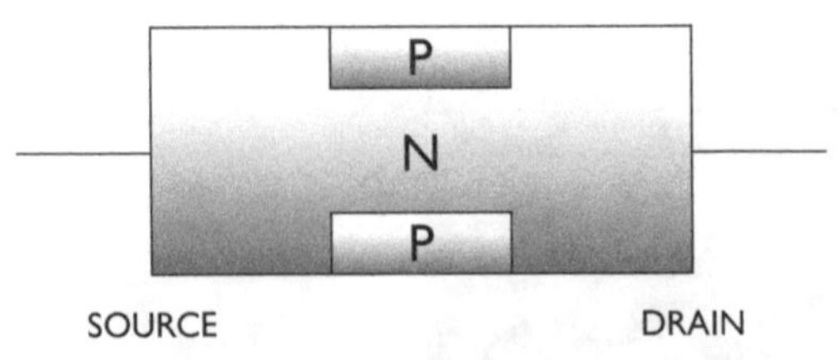

Fig. 2-8 Field-effect transistor.

Zener Diodes

The Zener diode is a PN diode that has been specially doped. Zener diodes are usually connected in circuits in the reverse-biased position and are used as surge protectors. Typically, they are installed parallel with a load that is to be protected, in the same manner as a lightning arrestor.

Connected in this way, Zener diodes oppose current flow (the definition of reverse-biased). But when the voltage applied to them reaches a certain level, called the *breakdown voltage,* they will conduct a current easily. This has the effect of shunting the voltage away from the load being protected and sending it through the Zener diode instead.

When properly sized Zener diodes are used this way, they provide a high level of overcurrent protection for sensitive circuits. They are especially useful because they have a very fast response time. The Zener diode will respond to an overvoltage within a few nanoseconds, rather than taking the many milliseconds of response time required by other types of surge suppressors before they can protect the circuit.

Working with Electronic Components

The basic rules of working with electrical components apply also to electronic devices: Handle with care, and make sure that you use parts at or below their rated voltage and wattage.

Most electronic parts are very durable and thus not often damaged by normal treatment. Nevertheless, you may want to pay a little extra attention to the temperatures at which they are stored or operated. High temperatures can have a deteriorating effect on certain electronic items. Also beware of installing parts with pins. Take care not to bend the pins; insert them straight into their places and don't twist or turn them. They simply can't take the stress.

Voltage and wattage ratings are critical. You must keep all items within their limits. Failure to do so will usually result in an instant problem. Although electronic parts can be extremely effective, they are not at all forgiving. They will promptly blow out if you apply them incorrectly.

If you are going to work with electronics, you will need to master one mechanical skill—soldering.

Fortunately, soldering is quite easy to do; you merely need to spend some time practicing. Get a good grade of soldering iron (properly called a soldering "pencil"), some rosin-core solder, and an old circuit board to practice on. The soldering pencil should be rated between 25 and 40 watts for electronics work. Too much wattage results in too much heat, which can damage some items. We won't take the space here to go through all the details of soldering. A good soldering iron should come with soldering instructions. Sorry, there are no shortcuts. You simply must practice until you have a good feel for what you are doing.

You may also want to get a desoldering tool. Desoldering tools are often necessary for removing components from circuit boards. As with the soldering iron, practice until you get it right.

Printed Circuit Boards

There are two main concerns when working with printed circuit boards. The first is that you install and remove them properly. They should always be inserted and/or removed with an end-to-end motion rather than with a side-to-side motion. See Fig. 2-9.

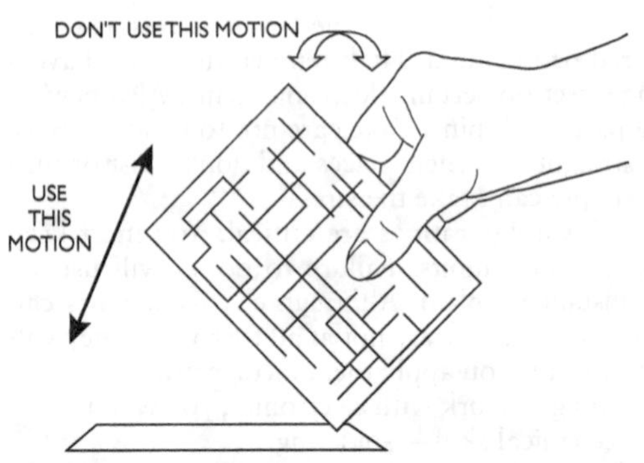

Fig. 2-9 Proper method of removing circuit boards.

The second concern with circuit boards is handling repairs or replacements. Because of their complexity, many of the components on these boards are nearly impossible to troubleshoot. In addition, manufacturers generally replace the entire board if you return it to them. But once a board has been worked on, the manufacturer has no way of knowing if the board was damaged because of a manufacturing error or because of your work on it. In these cases, manufacturers don't replace the board without payment. Call the manufacturer before you tamper with its boards, especially if they are still under warranty. Treat these boards with care; they are often worth hundreds or thousands of dollars.

Electronic Installations

Electronic systems require more preplanning than regular electrical systems. For an electrical project, you can grab several boxes of switches and receptacles and just wire them up. For electronic installations, you must know exactly where every item is supposed to go. These items are not "mix and match." In most electrical installations, no one will ever know or care if you use Hubbel wiring devices on one side of the building and Leviton on the other. But if you try mixing manufacturers in electronics, your chances are very slim that the system will work well. And even if it does, the manufacturers are often free of their warranty obligations if they find that you mixed brands of equipment.

This means extra planning is needed before the job begins. It also means that you need better working drawings and that installers need to pay more attention to details. You must use the exact devices and exact cable types designed for the installation. These systems can do amazing things, but they must be installed according to precise designs.

Also different from electrical installations is the way problems are solved. Everyone wants to solve problems with a minimum of hassle, but you must use extra caution in solving electronic problems. Unless you understand the system as well as the designer does, call the designer rather than flying on your own. If you don't, you could very easily fix the problem you are immediately concerned with but inadvertently create a larger problem in the process. All of the system's components must operate together; changing one piece can often affect several others.

Remember that you will probably have to spend some time fixing bugs in a complex system. Unless you are very familiar with a certain system, budget some extra time to fix small problems after the system becomes operational.

Testing

Plan on testing your electronic systems periodically as your work progresses. Again, this procedure is different from that used with regular electrical work, which is sometimes tested only just before the power is turned on, after the installation is complete.

Depending on what type of system you are installing, you may want to test several parts of the system before you even get close to turning it on.

With power or lighting circuits, the worst that can normally happen if you make a mistake is that a circuit breaker will snap at you. But with electronic systems, you can easily burn out thousands of dollars' worth of equipment.

These cautions are especially pertinent to electricians who learned power wiring carefully, but who are just beginning to understand electronic components and are doing so with little or no formal training. Proceed carefully, and don't take risks you don't completely understand. Err on the side of caution.

3. ELECTRICAL DRAWINGS

The construction documents supplied for a new building (normally by an architectural or engineering firm) include all architectural drawings that show the design and building construction details. These include floor plan layouts, vertical elevations of all building exteriors, various cross sections of the building, and other details of construction. While there may be a number of such drawings, they fall into five general groups:

1. **Site plans.** These plans include the location of the building and show the location and routing of all outside utilities (water, gas, electricity, sewer, etc.) that will serve the building, as well as other points of usage within established property lines. Topography lines are sometimes included with site plans, especially when the building site is on a slope.
2. **Architectural.** These drawings include elevations of all exterior faces of the building; floor plans showing walls, doors, windows, and partitions on each floor; and sufficient crosssections to clearly indicate various floor levels and details of the foundation, walls, floors, ceilings, and roof construction. Large-scale detail drawings may also be included.
3. **Structural.** Structural drawings are included for reinforced-concrete and structural-steel buildings. Structural engineers prepare these drawings.
4. **Mechanical.** The mechanical drawings cover the complete design and layout of the plumbing, piping, heating, ventilating, and air conditioning systems and related mechanical construction. Electrical control wiring diagrams for the heating and cooling systems are often included on the mechanical drawings as well.

5. **Electrical.** The electrical drawings cover the complete design and layout of the electrical wiring for lighting, power, signals and communications, special electrical systems, and related electrical work. These drawings sometimes include a site plan showing the location of the building (on the property) and the interconnecting electrical systems. They can also include floor plans showing the location of power outlets, lighting fixtures, panelboards, power-riser diagrams, and larger-scale details where necessary.

In order to read any of these drawings, you need to become familiar with the meanings of the many symbols, lines, and abbreviations used.

Plan Symbols

Because electrical drawings must be prepared by electrical draftsmen quickly and within budget, symbols are used to simplify the work. Therefore, anyone who must interpret and work with the drawings must have a solid knowledge of electrical symbols.

Most engineers and designers use electrical symbols adopted by the American National Standards Institute (ANSI). Many of these symbols, however, are frequently modified to suit a specific need for which there is no standard symbol. For this reason, most drawings include a symbol list or legend as part of the drawings or in the written specifications.

A listing of the most common types of plan symbols is shown in Table 3-1. This list represents a good set of electrical symbols in that they are (1) easily drawn by draftsmen, (2) easily interpreted by workmen, and (3) sufficient for most applications.

It is evident from Table 3-1 that many symbols have the same basic form, but their meanings differ slightly with the addition of a line, mark, or abbreviation. Therefore, a good procedure to follow in learning the different electrical symbols is first to understand the basic forms and then to apply the variations of that form to obtain the different meanings.

Table 3-1 Plan Symbols

Receptacle Outlets[a]	
	Single receptacle outlet.
	Duplex receptacle outlet.
	Triplex receptacle outlet.
	Quadruplex receptacle outlet.
	Duplex receptacle outlet—split wired.
	Triplex receptacle outlet—split wired.
*	Single special-purpose receptacle outlet[a].
*	Duplex special-purpose receptacle outlet[a].
R	Range outlet.
DW	Special-purpose connection or provision for connection. Use subscript letters to indicate function (DW—dishwasher; CD—clothes dryer, etc.).
X"	Multioutlet assembly. Extend arrows to limit of installation. Use appropriate symbol to indicate type of outlet. Also indicate spacing of outlets as *x* inches.
C	Clock hanger receptacle.
F	Fan hanger receptacle.
	Floor single receptacle outlet.

(continued)

Table 3-1 (continued)

Receptacle Outlets[a]	
	Floor duplex receptacle outlet.
	Floor special-purpose outlet[a].
	Floor telephone outlet—public.
	Floor telephone outlet—private.
Switch Outlets	
S	Single-pole switch.
S2	Double-pole switch.
S3	Three-way switch.
S4	Four-way switch.
SK	Key-operated switch.
SP	Switch and pilot lamp.
SL	Switch for low-voltage switching system.
SLM	Master switch for low-voltage switching system.
S	Switch and single receptacle.
S	Switch and double receptacle.
SD	Door switch.
ST	Time switch.
SCB	Circuit-breaker switch.
SMC	Momentary contact switch or push-button for other than signaling system.

Circuiting[c]	
	Wiring concealed in ceiling or wall.
	Wiring concealed in floor.
	Wiring exposed.
	Note: Use heavyweight line to identify service and feeders. Indicate empty conduit by notation CO (conduit only).
3 wires 2 1	Branch-circuit home run to panelboard. Number of arrows indicates number of circuits. (A numeral at each arrow may be used to identify circuit number.) Note: Any circuit without further identification indicates two-wire circuit. For a greater number of wires, indicate with cross lines.
4 wires, etc.	Unless indicated otherwise, the wire size of the circuit is the minimum size required by the specification.
	Identify different functions of wiring system—for example, signaling system by notation or other means.
	Wiring turned up.
	Wiring turned down.

Lighting Outlets		
Ceiling	***Wall***	
		Surface or pendant incandescent, mercury vapor, or similar lamp fixture.
R	R	Recessed incandescent, mercury vapor, or similar lamp fixture.

(continued)

Table 3-1 (continued)

Lighting Outlets		
Ceiling	***Wall***	
O		Surface or pendant individual fluorescent fixture.
OR		Recessed individual fluorescent fixture.
O		Surface or pendant continuous-row fluorescent fixture.
OR		Recessed continuous-row fluorescent fixture[a].
		Bare-lamp fluorescent strip[b].
X	X	Surface or pendant exit light.
XR	XR	Recessed exit light.
B	B	Blanked outlet.
J	J	Junction box.
L	L	Outlet controlled by low-voltage switching when relay is installed in outlet box.

Panelboards, Switchboards, and Related Equipment	
	Flush-mounted panelboard and cabinet[a].
	Surface-mounted panelboard and cabinet[a].
	Switchboard, power control center, unit substations[a] should be drawn to scale.
TC	Flush-mounted terminal cabinet.[a] In small-scale drawings the TC may be indicated alongside the symbol.
TC	Surface-mounted terminal cabinet.[a] In small-scale drawings the TC may be indicated alongside the symbol.

Panelboards, Switchboards, and Related Equipment	
	Pull box (identify in relation to wiring section and sizes).
	Motor or other power controller[a].
	Externally operated disconnection switch[a].
	Combination controller and disconnection means[a].

[a]Unless noted to the contrary, it should be assumed that every receptacle will be grounded and will have a separate grounding contact.

[b]Use the uppercase subscript letters described under Section 2 item a-2 of this Standard when weatherproof, explosion-proof, or some other specific type of device will be required.

Note also that some of the symbols listed contain abbreviations, such as WP for weatherproof and S for switch. Others are simplified pictographs, such as the symbols for a safety switch or panelboard. In other cases, the symbols are combinations of abbreviations and pictographs, such as the symbol for nonfusible safety switches.

Types of Electrical Drawings

The most common types of electrical drawings are these:

1. Electrical construction drawings
2. Single-line block diagrams
3. Schematic wiring diagrams

Electrical construction drawings show the physical arrangement and views of specific electrical equipment. These drawings give all the plan views, elevation views, and other details necessary to construct the installation. For example, Fig. 3-1 shows a pictorial sketch of a wire trough (auxiliary gutter). One side of the trough is labeled "top," one is labeled "front," and another is labeled "end."

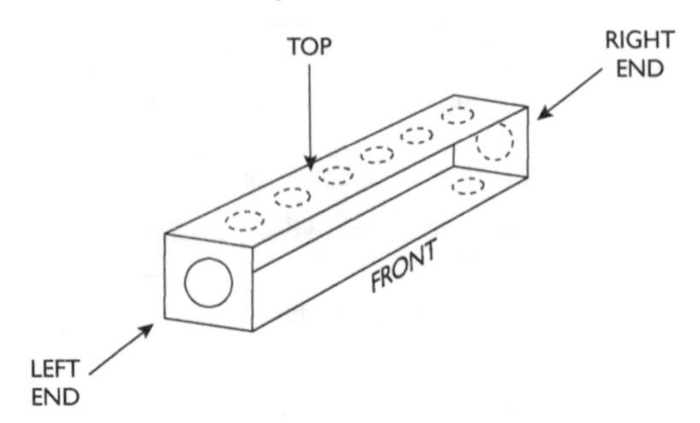

Fig. 3-1 Pictorial sketch.

This same trough is represented in another form in Fig. 3-2. The drawing labeled "top" is what one sees when viewing the panelboard directly from above; the one labeled "end" is viewed from the side; and the drawing labeled "front" shows the panelboard when viewing the panel directly from the front.

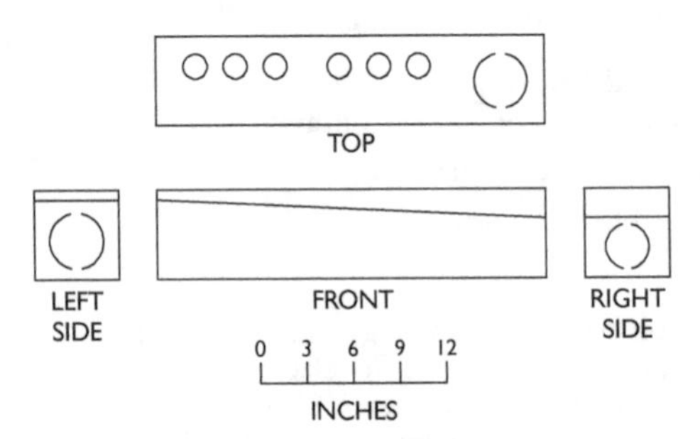

Fig. 3-2 Top, front, and side views.

The width of the trough is shown by the horizontal lines of the top view and the horizontal lines of the front view. The height is shown by the vertical lines of both the front and the end views, while the depth is shown by the vertical lines of the top views and the horizontal lines of the side view.

The three drawings in Fig. 3-2 clearly give the shape of the wire trough, but the drawings alone would not enable a worker to construct it, because there is no indication of the size of the trough. There are two common methods to indicate the actual length, width, and height of the wire trough. The first is to draw all of the fields to some given scale, such as 1½ in. = 1ft 0 in. This means that 1½ in. on the drawing represents 1 ft in the actual construction. The second method is to give dimensions on the drawings like the one shown in Fig. 3-3. Note that the gauge and type of material are also given in this drawing; there is enough data to show clearly how the panelboard is to be constructed.

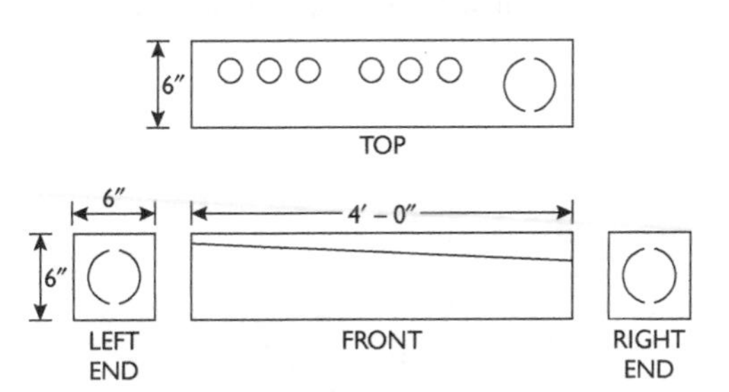

Fig. 3-3 Alternate-method top, front, side views.

Electrical construction drawings like the ones just described are used mainly by electrical equipment manufacturers. The electrical installer will more often run across electrical construction drawings like the one shown in Fig. 3-4. This type of construction drawing is normally used to supplement a building's electrical system drawings for a special installation and is often referred to as an electrical detail drawing.

Electrical diagrams intend to show, in diagrammatic form, electrical components and their related connections. In diagrams, electrical symbols are used extensively to represent the various components. Lines are used to connect these symbols, indicating the size, type, and number of wires that are necessary to complete the electrical circuit.

The electrical contractor will often come into contact with *single-line block diagrams*. These diagrams are used extensively to indicate the arrangement of electrical services on electrical working drawings. The power-riser diagram in Fig. 3-5 is typical of such drawings. It shows the panelboard and all related equipment, as well as the connecting lines that indicate the circuits and feeders. Notes are used to identify each piece of equipment and to indicate the size of conduit necessary for each circuit or feeder, as well as the number, size, and type of insulation on the conductors in each conduit.

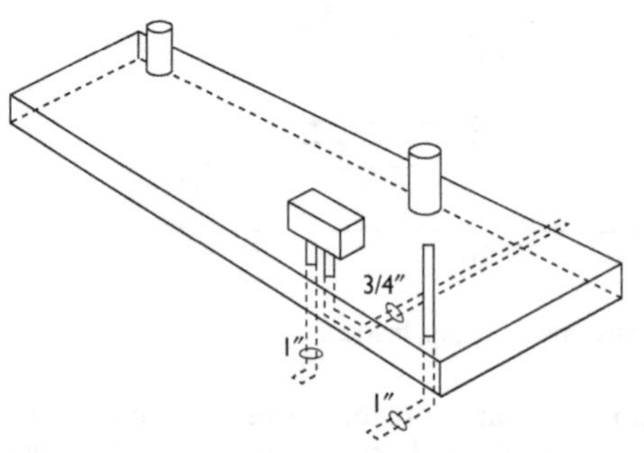

Fig. 3-4 Perspective drawing.

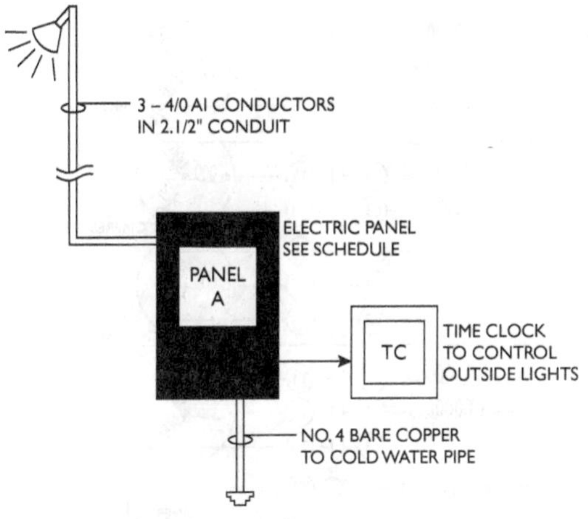

Fig. 3-5 Power-riser diagram.

A *schematic wiring diagram,* as shown in Fig. 3-6, is similar to a single-line block diagram except that the schematic diagram gives more detailed information and shows the actual size and number of wires used for the electrical connections.

Anyone involved in the electrical construction industry, in any capacity, frequently encounters all three types of electrical drawings. Therefore, it is important for everyone involved in this industry to fully understand electrical drawings, wiring diagrams, and other supplementary information found in working drawings and in written specifications.

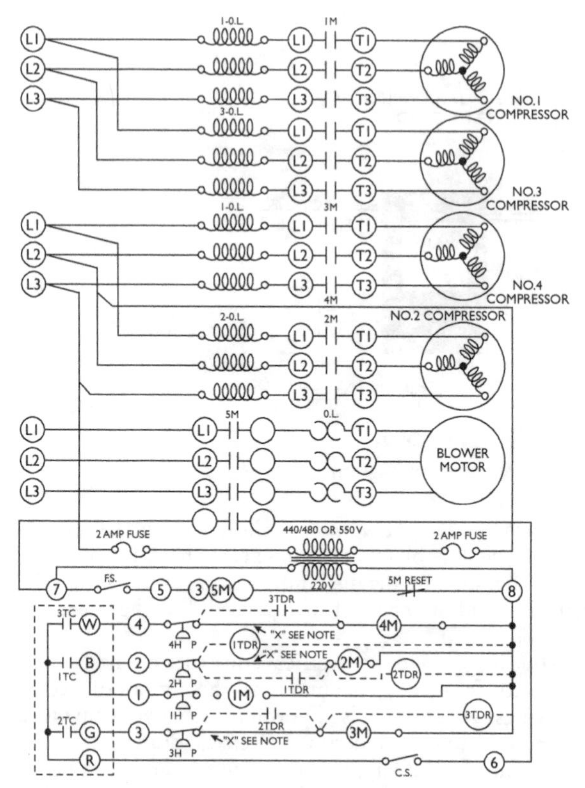

Fig. 3-6 Schematic wiring diagram.

Wiring Diagrams

Complete schematic wiring diagrams are used infrequently on the average set of electrical working drawings (only for complicated systems such as control circuits), but it is important to have a thorough understanding of them when the need to interpret them arises.

Components in schematic wiring diagrams are represented by symbols, and every wire is shown either by itself or included in an assembly of several wires that appear as one line on the drawing. Each wire in the assembly is numbered when it enters, however, and it keeps the same number when it emerges to be connected to some electrical component in the system. Fig. 3-7 shows a complete schematic wiring diagram for a three-phase ac magnetic motor starter. Note that this diagram shows the various devices (in symbol form) and indicates the actual connections of all wires between the devices.

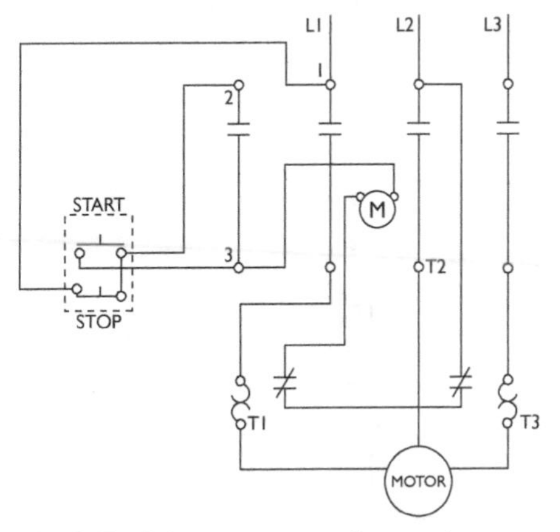

Fig. 3-7 Schematic wiring diagram for magnetic motor starter.

Fig. 3-8 gives a list of electrical wiring symbols commonly used for *single-line schematic* diagrams. Single-line diagrams are simplified versions of complete schematic diagrams. Fig. 3-9 shows the use of these symbols in a typical single-line diagram of an industrial power-distribution system.

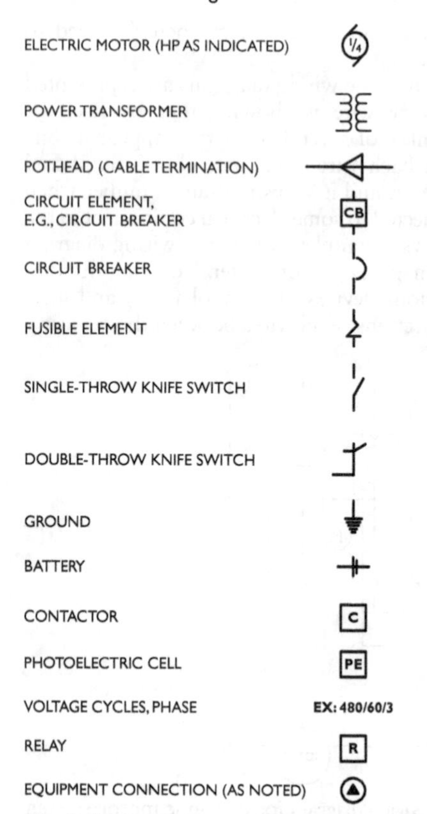

Fig. 3-8 Symbols used for single-line schematics.

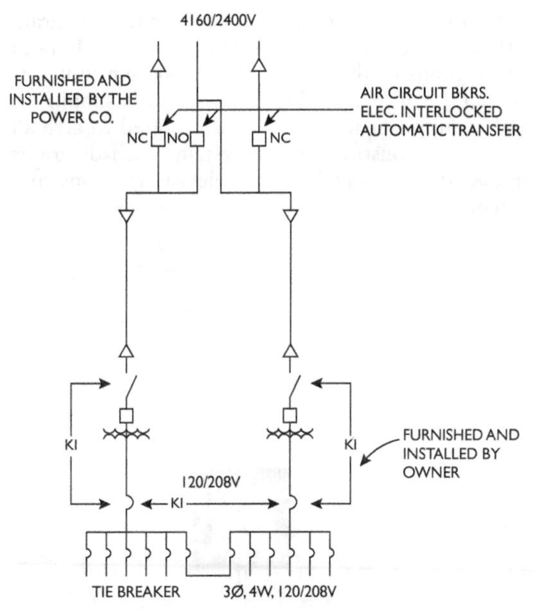

Fig. 3-9 Diagram of industrial power-distribution system.

Power-riser diagrams are probably the most frequently encountered diagrams on electrical working drawings for building construction. Such diagrams give a picture of what components are to be used and how they are to be connected in relation to one another. This type of diagram is more easily understood at a glance than diagrams previously described. As an example, compare the power-riser diagram in Fig. 3-10 with the schematic diagram in Fig. 3-11. Both are wiring diagrams of the same electrical system, but it is easy to see that the drawing in Fig. 3-10 is greatly simplified, even though a supplemental schedule is required to give all necessary data for constructing the system. Such diagrams are also frequently used on telephone, television, alarms, and similar systems.

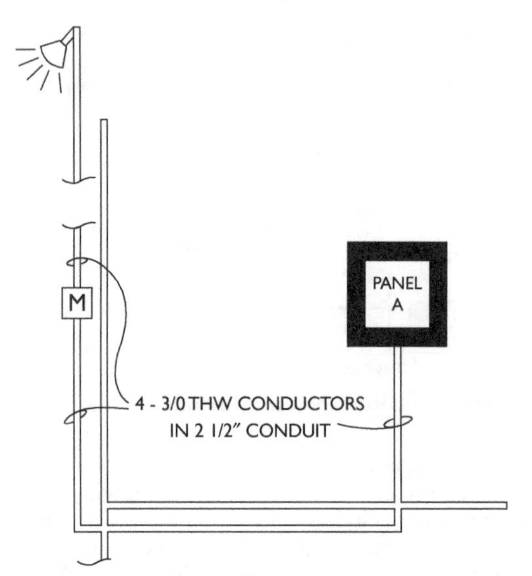

Fig. 3-10 Power-riser diagram.

Site Plans

A site plan is a plan view that shows the entire property, with the buildings drawn in their proper locations on the plot. Such plans sometimes include sidewalks, driveways, streets, and utility systems related to the building or project.

Site plans are drawn to scale using the engineer's scale rather than the architect's scale used for most building plans. On small lots, a scale of 1 in. = 10 ft or 1 in. = 20 ft is commonly used. This means that 1 in. (actual measurement on the drawing) is equal to 10 ft or 20 ft on the land itself.

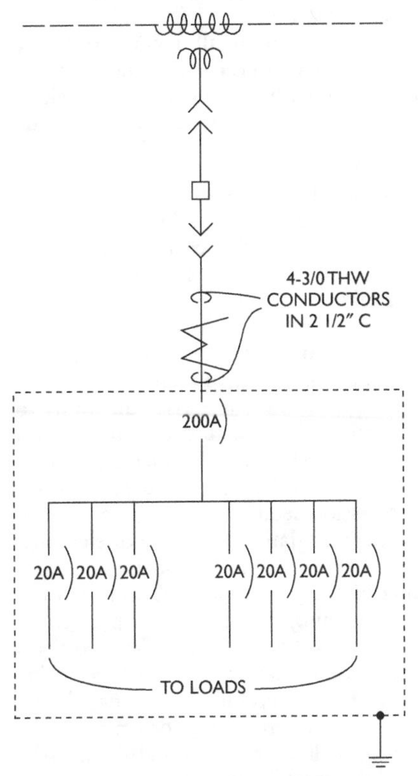

Fig. 3-11 Schematic diagram of the same system as Fig. 3-10.

In general building construction practice, it is the owner's responsibility to furnish the architect with property and topographic surveys made by a certified land surveyor or civil engineer. These surveys will show (1) all property lines, (2) existing utilities and their location on or near the property, (3) the direction of the land slope, and (4) the condition of the land (rocky, wet, or whatever).

The site plan is used to incorporate all new utilities. The electrical installer will then be concerned with the electrical distribution lines, the telephone lines, and the cable television lines, especially if they are to be installed underground.

Layout of Electrical Drawings

The ideal electrical drawing should show in a clear, concise manner exactly what is required of the installer. The amount of data shown on such drawings should be sufficient, but not overdone. Unfortunately, this is not always the case. The quality of electrical drawings varies widely.

In general, a good set of electrical drawings should contain floor plans for each floor of the building (assuming that the project is a building), including one plan for lighting circuitry and one plan for power circuitry; riser diagrams to show the service equipment, feeders, and communication equipment diagrammatically; schedules to indicate the components of the service equipment, lighting fixtures, and similar equipment; and large-scale detailed drawings for special or unusual portions of the installation. A legend or electrical symbol list should also be provided on the drawings in order to explain the meaning of every symbol, line, and notation used on the drawings. Anything that can't be explained by symbols and lines should be clarified with neatly lettered notes or explained in the written specifications. The scale to which the drawings are prepared is also important. Drawings should be as large as practical, and where dimensions need to be extremely accurate, dimension lines should be added. Fig. 3-12 shows a poorly prepared electrical drawing, whereas Fig. 3-13 shows one of relatively good quality. In Fig. 3-12 it is obvious that the electrical contractor will have to lay out or design portions of the system before it can be properly installed.

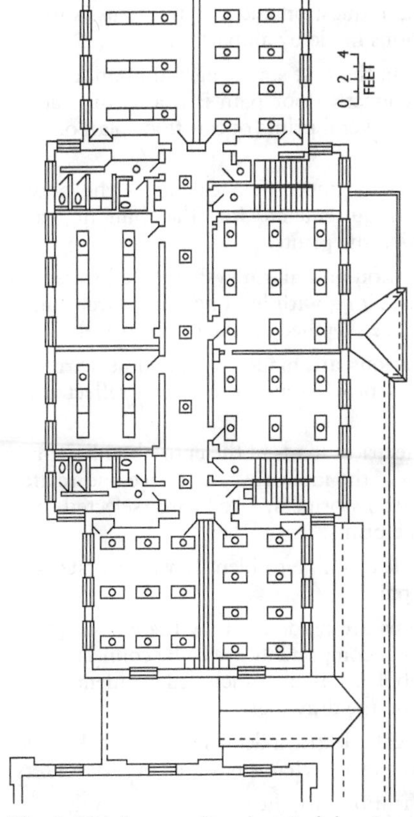

Fig. 3-12 Low-quality electrical drawings.

The following steps are necessary in preparing a good set of electrical working drawings and specifications:

1. The engineer or electrical designer meets with the architect and the owner to discuss the electrical needs of the building in question and also to discuss various recommendations made by all parties.
2. Once the data in the first step is agreed upon, an outline of the architect's floor plan is drawn on tracing paper and then several prints of this floor plan outline are made.
3. The designer or engineer then calculates the power and lighting requirements for the building and sketches them on the prints.
4. All communication and alarm systems are located on the floor plans, along with lighting and power panelboards. These are sketched on the prints as well.
5. Circuit calculations are made to determine wire size and overcurrent protection and are then reflected on the drawings.
6. After all the electrical loads in the entire building have been determined, the main electric service and related components (transformers, etc.) are selected and sketched on the prints.
7. Schedules are next in line to identify various pieces of electrical equipment.
8. Wiring diagrams are made to show the workers how various electrical components are to be connected. An electrical symbol list is also included to identify the symbols used on the drawings.
9. Various large-scale electrical details are included, if necessary, to show exactly what is required of the workers.
10. Written specifications are then made to give a description of the materials and the installation methods.

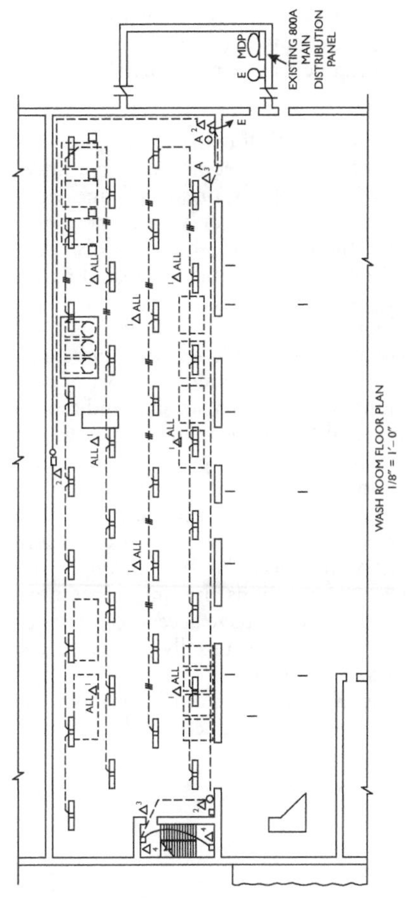

Fig. 3-13 Good-quality electrical drawings.

If these steps are properly taken in preparing a set of electrical working drawings, the drawings will be sufficiently detailed and accurate to enable a more rapid installation.

Schedules

A schedule, as related to electrical drawings, is a systematic method of presenting notes or lists of equipment on a drawing, in tabular form. When properly organized and thoroughly understood, schedules not only are powerful, time-saving methods for the draftsmen, but they also save the electrical personnel much valuable time in installing the equipment in the field.

For example, the lighting fixture schedule in Fig. 3-14 lists the fixture type corresponding to letters or numbers on the drawings. The manufacturer and catalog number of each fixture are included, along with the number, size, and type of lamp for each. The Volts and Mounting columns follow, and the column on the extreme right is for special remarks such as the mounting height for a wall-mounted fixture.

Sometimes schedules are omitted from the drawings, and the information is placed in the written specifications instead. This is not a good practice. Combing through page after page of written specifications is time-consuming. Furthermore, workers don't always have access to the specifications while working, whereas they usually do have access to the working drawings at all times.

The schedules in Figs. 3-15 through 3-17 are typical of those used by consulting engineers on electrical drawings.

LIGHTING FIXTURE SCHEDULE						
FIXT. TYPE	MANUFACTURER'S DESCRIPTION	LAMPS		VOLTS	MOUNTING	REMARKS
		NO.	TYPE			

Fig. 3-14 Lighting fixture schedule.

FRANK J. SULLIVAN ASSOCIATES
CONSULTING ENGINEERS
WASHINGTON 36, D.C.

PROJECT: ____________ DATE: ________

JOB: ______________ BY: __________

INTERCOMMUNICATION SYSTEM SCHEDULE																													
ROOM NAME	ITEM	I.D. NO.	1	2	3	4	5	6	7	8	9	10	11	12	13	14	15	16	17	18	19	20	21	22	23	24	25	26	REMARKS
		1																											
		2																											
		3																											
		4																											
		5																											
		6																											
		7																											
		15																											
		16																											
		17																											
		18																											
		19																											
		20																											
		21																											
		22																											
		23																											
		24																											
		25																											
		26																											

Fig. 3-15 Intercom schedule.

LOAD CENTER UNIT SUBSTATION EQUIPMENT SCHEDULE					
	SWITCH				
ITEM	SWITCH RATING	POLES	FUSE RATING	EQUIPMENT	DESIGNATION

Fig. 3-16 Load center schedule.

DATE ______ PROJECT/JOB NO. ______________ MAINS ___ A.
BY ______ MOTOR CONTROL CENTER NO. ___ MCC ___ ___Ø ___ V.

ITEM NO	HP KW	FLA	STR. SIZE	P'S	SW SIZE	FUSE SIZE	AUX EQUIP	REM CONTROL DEVICES	POWER CONTROL DIAGRAM	NAMEPLATE DESIGNATION	REMARKS
											__MCC__

Fig. 3-17 Motor control schedule.

Sectional Views

Sometimes the construction of a building is difficult to show with the regular projection views normally used on electrical drawings. For example, if too many broken lines are needed to show hidden objects in buildings or equipment, the drawings become confusing and difficult to read. Therefore, in most cases building sections are shown on working drawings, to clarify the construction. To better understand a building section, imagine that the building has been cut into sections with a saw. The floor plan of a building in Fig. 3-18 shows a sectional cut at point A-A. This sectional view is then shown in Fig. 3-19.

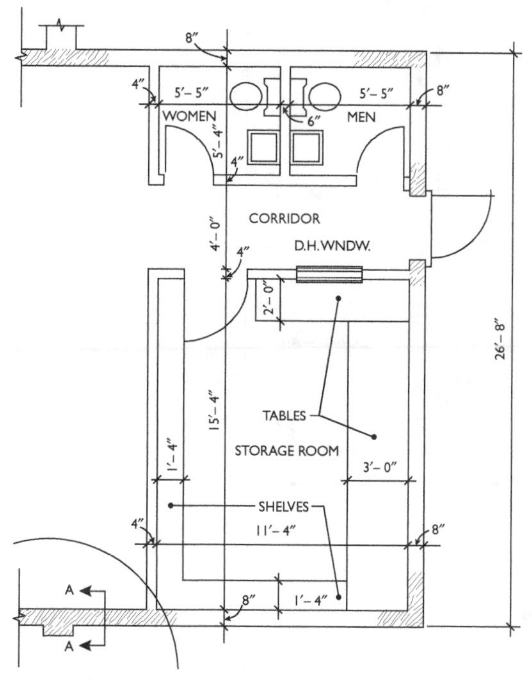

Fig. 3-18 Floor plan showing cut at A-A.

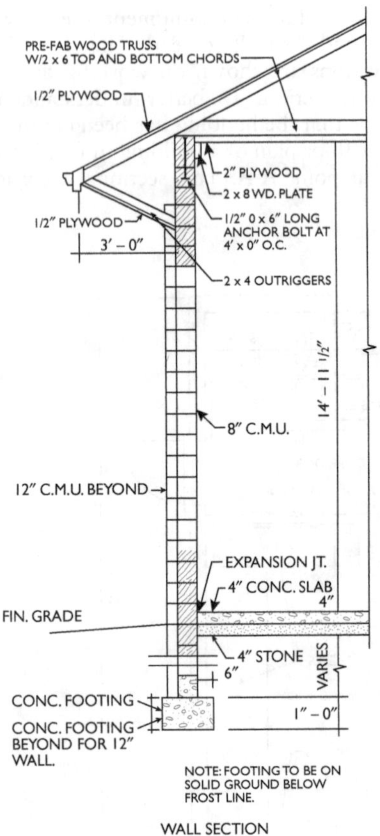

Fig. 3-19 Sectional view of cut shown in Fig. 3-18.

Fig. 3-20 Cutting-plane line.

In dealing with sections, it is important to use a considerable amount of visualization. Some sections are very easy to read and others are extremely difficult, because there are no set rules for determining what a section will look like. For example, a piece of rigid conduit, cut vertically, will have the shape of a rectangle; cut horizontally, it will have a round or circular appearance; but on the slant, it will be an ellipse.

A cutting-plane line (Fig. 3-20) has arrowheads to show the direction in which the section is viewed. Letters such as A-A or B-B are normally used with the cutting-plane lines to identify the cutting plane and the corresponding sectional views.

4. MOTORS, CONTROLLERS, AND CIRCUITS

Motors are among the most commonly used electrical devices. They vary in size, from specially designed medical motors that are less than an inch long to gigantic industrial units of several thousand horsepower. In between are literally hundreds of different types of motors for thousands of different applications. Therefore, it is very important for the electrical installer to understand the rules regarding the application and wiring of motors. Generally, the concerns can be broken down into four categories:

1. **Mechanical safety.** You must ensure that the motors themselves don't constitute a source of danger. For instance, you must take care not to install open motors in areas where they attract curious children to investigate and injure themselves. Likewise, it is often a good idea to put a clutch on a motor to avoid possible injury to the machine operator.
2. **Mechanical stability and operations.** Motors have a number of mechanical stresses placed upon them. One of the primary forces is vibration, which has the unfortunate side effect of loosening bolts and screws. Vibration causes mechanical difficulties to the motor and to the equipment that it operates. Surrounding items may also be affected.
3. **Electrical safety.** The first issue here is to make sure that motors don't become the source of an electrical shock or fault. Another issue is to make sure that motors don't cause problems to the electrical system on which they are installed.
4. **Operational circuits.** One last concern is that the circuits on which motors are installed can operate continually and correctly. Motors place unusual demands on electrical circuits. First of all, they can require large

> starting currents. (Fully loaded motors can draw starting currents of 4 to 8 times their normal full-load current—in some circumstances even higher.) They also put a lot of inductive reactance into electrical systems. And because of the high currents that some motors draw, they overheat electrical circuits more readily than do many other types of loads.

It is the responsibility of the installer to ensure that the equipment installed (in this case, electric motors) is safe for the end user.

The Basics

The operation of electric motors involves not only current and voltage but also magnetic fields and their associated characteristics.

Basically, all electric motors operate by using electromagnetic induction, which, simply put, is the interaction between conductors, currents, and magnetic fields. Any time an electrical current passes through a conductor (of which copper wires are the most common type), it causes a magnetic field to form around that conductor. This is one of the absolute laws of physics. Conversely, any time a magnetic field moves through a conductor, it induces (causes to flow) an electrical current in that conductor. Again, this is an absolute and unchangeable law of physics.

The manipulation of these two laws, in combination with the principles of magnetic attraction and repulsion, is the basis for the operation of motors. The intelligent use of electromagnetic induction turns electricity into physical force, thereby causing the motor to turn.

Let us go just one step further and explain in a little more detail how this occurs. These are the basic operations of an electric motor, step by step: An electrical current is turned on and flows through the motor's windings, causing a strong magnetic field to form around the windings. This magnetic field attracts the rotor (the part in the center of the motor that turns the shaft is at the center of the rotor) and moves it toward the magnetic field, causing the initial movement of the motor. This movement is perpetuated by any one of various means of rotating the magnetic field. The most common method is to use several different windings to which current is sent alternately, thus causing magnetic strength to be in one place one moment and another place the next. The rotor then follows these fields, creating continuous motion.

Although there are any number of variations and modifications to these basic operations, these are the principles by which all motors function. Depending on the type of motor design, you can increase or decrease power, operate at different voltages, and control motor speed.

Motor Mountings

In general, motors must be installed so that adequate ventilation is provided and maintenance operations can be performed without difficulty.

Open motors (motors whose windings are not fully enclosed) with commutators or collector rings must be located so that sparks from the motors cannot reach combustible materials. However, this does not prohibit installation of these motors on wooden floors.

Suitably enclosed motors must be used in areas where significant amounts of dust are present.

One of the most important considerations for the mounting of electric motors is that they be installed so that vibration won't be a problem. Generally, this requires a careful mounting using strong fasteners. If you suspect that there will be trouble with vibration, there are special antivibration mountings that you can use to minimize the difficulty.

When fastening to a concrete base, the ideal method is to have J-bolts installed in the concrete pour. In such a case, templates must be carefully prepared prior to the pour. Placement of the templates must be double-checked on the day the concrete is poured into place.

Consider using a large-size lag shield or a lead anchor in pre-existing concrete. Maximum strength is of utmost importance in these cases. If the anchor loosens, vibration of the motor will increase dramatically. This, of course, will cause the fastener to loosen further, and the motor will experience serious difficulties in a fairly short time.

When anchoring to a metal base, the base should be drilled and tapped and the motor fastened to it with machine screws and lock washers.

When anchoring to a wood base, it is generally preferable to drill entirely through the wood and fasten the motor with carriage bolts, fender washers, and lock washers. For small motors on strong wood bases, you can use a very large wood or sheet metal screw.

Maintenance

With all motors, it is necessary to provide regular maintenance. Even though most modern ac motors have no need of lubrication or changing of brushes as did older motors, they do need to be periodically checked for problems. The chief problems to look for are vibration, overheating, and proper alignment of any pulleys, gears, or belts.

Fig. 4-1 shows a commonly used type of motor maintenance schedule that can be used to keep track of maintenance operations. It lists the motor number and location, basic operating characteristics, and service data for a number of motors.

Some maintenance mechanics prefer to keep a separate card file, with an index card for each motor. And in recent years, several types of computer programs have been developed. These programs provide an easy way to save data on a large number of motors. Any of these methods is effective, provided that you use it consistently.

In general, every electrical motor in an industrial setting should be carefully serviced at least once per month. Some older motors have grease fittings for lubrication; most newer motors don't. They should all be checked regularly, however, and be given whatever attention they need.

MOTOR DATA FOR													
MOTOR NO.	MACHINE DRIVEN	HP	VOLTAGE	PHASE	SERIAL NO.	FRAME	AMPS	RPM	CODE	MOTOR STARTER			
										MAKE	NEMA RATING	COIL NO.	STR SIZE

Fig. 4-1 Motor schedule.

Conductors for Motor Circuits

Proper sizing of motor conductors and overcurrent protection are the most important factors in a motor installation. Fig. 4-2 is a brief outline of the various steps to be taken in motor circuit design. Although these steps are explained in depth in this chapter, referring to a brief outline such as this one is generally the quickest way to determine circuit requirements.

Branch-circuit conductors that supply *single motors* must have an ampacity of at least 125 percent of the motor's full-load current rating. (You will find full-load current ratings in *Tables 430.147* through *430.150* of the National Electrical Code.) This is necessary because motors cause temporary surges of current that could overheat the conductors if they are not oversize.

Motors used only for short cycles can have their branch-circuit ampacities reduced according to *Table 430.22(E)*.

DC motors fed by single-phase rectifiers must have the ampacity of their conductors rated at 190 percent of the full-load current for half-wave systems and 150 percent of the full-load current for full-wave systems. (This is because of the high levels of current that these motors can draw from the rectifiers.)

DESIGNING MOTOR CIRCUITS

For One Motor:

1. Determine full-load current of motor(s) (*Table 430.150* for 3 phase).
2. Multiply full-load current × 1.25 to determine minimum conductor ampacity (*Section 430.22(A)*).
3. Determine wire size (*Table 310.16*).
4. Determine conduit size. (*Table C4*).
5. Determine minimum fuse or circuit breaker size (*Table 430.72(B)*) (*Section 240.6*).
6. Determine overload rating (*Section 430.32(C)*).

For more than one motor:

1. Perform steps 1 through 6 as shown above for each motor.
2. Add full-load current of all motors, plus 25% of the full-load current of the largest motor to determine minimum conductor ampacity (*Section 430.24*)
3. Determine wire size (*Table 310.16*).
4. Determine conduit size. (*Table C4*).
5. Add the fuse or circuit breaker size of the largest motor, plus the full-load currents of all other motors to determine the maximum fuse or circuit breaker size for the feeder (*Section 430.24*) (*Section 240.6*).

Fig. 4-2 Sizing of motor circuitry.

For phase converters, the single-phase conductors that supply the converter must have an ampacity of at least 2.16 times the full-load current of the motor or load being served. (This assumes that the voltages are equal. If they are not, the calculated current must be multiplied by the result of output voltage divided by input voltage.)

Conductors connecting secondaries of *continuous-duty wound-rotor motors* to their controllers must have an ampacity of at least 125 percent of the full-load secondary current.

When a resistor is installed separately from a controller, the ampacity of the conductors between the controller and the resistor must be sized according to *Table 430.23(C)*.

Conductors That Supply Several Motors or Phase Converters

Conductors that supply *two or more motors* must have an ampacity of no less than the total of the full-load currents of all motors being served, plus 25 percent of the highest rated motor in the group. If interlock circuitry guarantees that all motors can't be operated at the same time, the calculations can be made based on the largest group of motors that can be operated at any time.

Several motors can be connected to the same branch circuit if the following requirements are met:

1. Motors installed on general-purpose branch circuits without overload protection are motors of only 1 horsepower or less (assuming the installation complies with all other requirements).
2. The full-load current is not more than 6 amperes.
3. The branch-circuit protective device rating marked on any controllers is not exceeded.

Conductors supplying two or more motors must be provided with a protective device rated no greater than the highest rating of the protective device of any motor in the group *plus* the sum of the full-load currents of the other motors.

Where *heavy-capacity feeders* are to be installed for future expansions, the rating of the feeder protective devices can be based on the ampacity of the feeder conductors.

Phase converters must have an ampacity of 1.73 times the full-load current rating of all motors being served, plus 25 percent of the highest rated motor in the group. (This assumes that the voltages are equal. If they are not, the calculated current must be multiplied by the result of output voltage divided by input voltage.) If the ampere rating of the 3-phase output conductors is less than 58 percent of the rating of the single-phase input current ampacity, separate overcurrent protection must be provided within 10 ft of the phase converter.

Conductors That Supply Motors and Other Loads

Conductors that supply both motors and other loads must have their motor loads computed as specified above, other loads computed according to their specific Code requirements, and all loads then added together.

If *taps* are to be made from feeder conductors, they must terminate in a branch-circuit protective device and must:

1. Have the same ampacity as the feeder conductors;

   OR

2. Be enclosed by a raceway or in a controller, and be no longer than 25 ft;

   OR

3. Have an ampacity of at least one-third of the feeder ampacity, be protected, and be no longer than 25 ft.

A tap circuit is illustrated in Fig. 4-3.

In *high-bay manufacturing buildings* (more than 25 ft from floor to ceiling, measured at the walls), taps longer than 25 ft are permitted. In these cases:

1. Tap conductors must have an ampacity of at least one-third that of the feeder conductors.
2. Tap conductors must terminate in an appropriate circuit breaker or set of fuses.

3. Tap conductors must be protected from damage and be installed in a raceway.
4. Tap conductors must be continuous, with no splices.
5. The minimum size of tap conductors is No. 6 AWG copper or No. 4 AWG aluminum.
6. Tap conductors can't penetrate floors, walls, or ceilings.
7. Tap conductors may be run no more than 25 ft horizontally and no more than 100 ft overall.

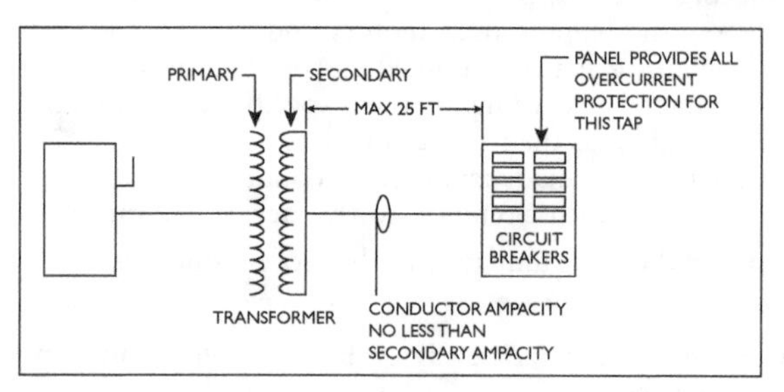

Fig. 4-3 Tap circuits not over 25 ft long.

Feeders that supply motors and lighting loads must be sized to carry the entire lighting load plus the motor load.

Grounding

As with virtually all other electrical equipment, motors need to be grounded to maintain safety. There are rare exceptions, but the requirement applies to almost all motors. Grounding requirements are generally as follows:

The frames of *portable motors* that operate at more than 150 volts (V) must be grounded or guarded.

The frames of stationary motors *must be grounded* (or permanently and effectively isolated from ground) in the following circumstances:

1. When supplied by metal-enclosed wiring.
2. In wet locations, when they are not isolated or guarded.
3. In hazardous locations.
4. If any terminal of the motor is more than 150 V to ground.

All *controller enclosures* must be grounded, except when attached to portable ungrounded equipment.

Controller-mounted devices must be grounded.

Overload Protection

Overload protection is not required where it might increase or cause a hazard, as would be the case if overload protection were used on fire pumps.

Continuous-duty motors of more than 1 horsepower must have overload protection. This protection may be in one of the following forms:

1. The motor may have an overload device that responds to motor current. Such units must be set to trip at 115 percent of the motor's full-load current (based on the nameplate rating). Motors with a service factor of at least 1.15 or with a marked temperature rise of no more than 40°C can have their overloads set to trip at 125 percent of the full-load current (see Fig. 4-4).
2. Also acceptable are any of several methods of required overload protection built into motors by the manufacturer (not by the installer).

Motors of *1 horsepower or less* that are not permanently installed, are manually started, and are within sight of their controller are considered to be protected from overload by their branch-circuit protective device. They may be installed on 120-volt circuits up to 20 amps.

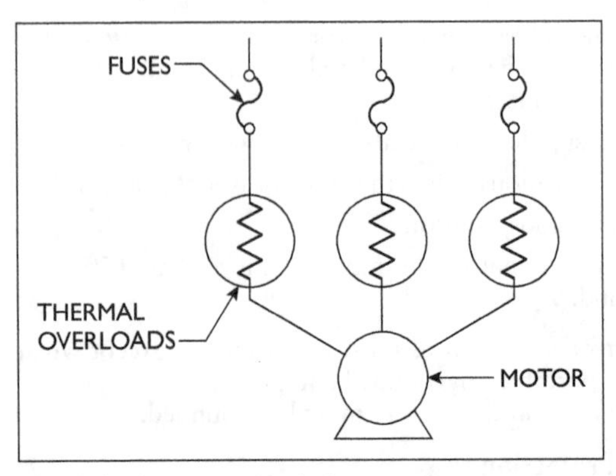

Fig. 4-4 Motor overload placement.

Motors of 1 horsepower or less that are permanently installed, automatically started, or not within sight of their controllers may be protected from overloads by one of the following methods:

1. The motor may have an overload device that responds to motor current. Such a unit must be set to trip at 115 percent of the motor's full load current (based on the nameplate rating). Motors with a service factor of at least 1.15 or with a marked temperature rise of no more than 40°C can have their overloads set to trip at 125 percent of the full-load current.
2. Also acceptable are any of several methods of required overload protection built into motors by the manufacturer (not by the installer).
3. Motors that have enough impedance to ensure that overheating is not a threat can be protected by their branch-circuit protective devices only.

Wound-rotor secondaries are considered to be protected from overload by the motor overload protection.

Intermittent-duty motors can be protected from overload by their branch-circuit protective devices only as long as the rating of the branch-circuit protective device does not exceed the rating specified in *Table 430.22(E)* of the NEC.

In cases where normal overload protection is too low to allow the motor to start, it may be increased to 130 percent of the motor's full-load rated current. Motors with a service factor of at least 1.15 or with a marked temperature rise of no more than 40°C can have their overloads set to trip at 140 percent of full-load current.

Manually started motors are allowed to have their overload protective devices momentarily cut out of the circuit during the starting period. The design of the cutout mechanism must ensure that it won't allow the overload protective devices to remain cut out of the circuit.

If *fuses* are used as overload protection, they must be installed in all ungrounded motor conductors as well as in the grounded conductor for 3-phase, 3-wire systems that have one grounded wire (corner-ground delta systems).

If *trip coils, relays, or thermal cutouts* are used as overload protective devices, they must be installed according to *Table 430.37* of the NEC. The requirements for standard motor types are as follows:

> *Three-phase ac motors:* One overload device must be placed in each phase.
>
> *Single-phase ac or dc, one wire grounded:* One overload device in the ungrounded conductor.
>
> *Single-phase ac or dc, ungrounded:* One overload device in either conductor.
>
> *Single-phase ac or dc, three wires, grounded neutral:* One overload device in either ungrounded conductor.

In general, overload protective devices should open enough ungrounded conductors to stop the operation of the motor.

When motors are installed on *general-purpose branch circuits,* their overload protection must be as follows:

1. Motors of more than 1 horsepower can be installed on general-purpose branch circuits only when their full-load current is less than 6 amperes, they have overload protection, the branch-circuit protective device rating marked on any controller is not exceeded, and the overload device is approved for group installation.
2. Motors of 1 horsepower or less can be installed on general-purpose branch circuits without overload protection (assuming the installation complies with the other requirements mentioned above) if the full-load current is not more than 6 amperes and the branch-circuit protective device rating marked on any controller is not exceeded.
3. When a motor is cord-and-plug connected, the rating of the plug and receptacle may not be greater than 15 amperes at 125 V or 10 amperes at 250 V. If the motor is more than 1 horsepower, the overload protection must be built into the motor. The branch circuit must be rated according to the rating of the cord and plug.
4. The branch and overload protection must have enough time delay to allow the motor to start.

Overload protective devices that can restart a motor automatically after tripping are not allowed unless approved for use with a specific motor. They are *never* allowed if their use can cause possible injury.

In cases where the instant shutdown of an overloaded motor would be dangerous to people (as could be the case in a variety of industrial settings), a supervised alarm may be used, followed by an orderly (rather than instant) shutdown.

Short-Circuit and Ground-Fault Protection

Short-circuit and ground-fault protective devices must be capable of carrying the starting current of the motors they protect.

In general, protective devices must have a rating of no less than the values given in *Table 430.52* of the NEC. When these values don't correspond with the standard ratings of overcurrent protective devices, the next higher setting can be used.

If the rating given in *Table 430.52* of the NEC is not sufficient to allow for the motor's starting current, the following methods can be used:

1. A non-time-delay fuse of 600 amps or less can be increased enough to handle the starting current, but not to more than 400 percent of the motor's full-load current.
2. A time-delay fuse can be increased enough to handle the starting current, but not to more than 225 percent of the motor's full-load current.
3. The rating of an inverse time circuit breaker can be increased, but not to more than 400 percent of full-load currents that are 100 amps or less, or 300 percent of full-load currents over 100 amps.
4. The rating of an instantaneous trip circuit breaker can be increased, but not to more than 1300 percent of the full-load current.
5. Fuses rated between 601 and 6000 amps can be increased, but not to more than 300 percent of the rated full-load current.

Instantaneous trip circuit breakers can be used as protective devices, but only if they are adjustable and are part of a listed combination controller that has overload, short-circuit, and ground-fault protection in each conductor.

A motor's *short-circuit protector* can be used as a protective device, but only when it is part of a listed combination controller that has overload, short-circuit, and ground-fault protection in each conductor and does not operate at more than 1300 percent of full-load current.

For *multispeed motors,* a single short-circuit and ground-fault protective device can protect two motor windings as long as the rating of the protective device is not greater than the highest possible rating for the smallest winding. (Multipliers of *Table 430.52* of the NEC are used for the rating of the smallest winding.)

A single *short-circuit and ground-fault protective device* can be used for multispeed motors, sized according to the full-load current of the highest rated winding. However, each winding must have its own overload protection, which must be sized according to each winding. Also, the branch-circuit conductors feeding each winding must be sized according to the full-load current of the highest winding.

If branch circuit and ground-fault protection ratings are shown on motors or controllers, they must be followed even if they are lower than Code requirements.

Fuses can be used instead of the devices mentioned in *Table 430.52* of the NEC for adjustable-speed drive systems, as long as a marking for replacements is provided next to the fuse holders.

For *torque motors,* the *branch-circuit protection* must be equal to the full-load current of the motor. If the full-load current is 800 amps or less and the rating does not correspond to a standard overcurrent protective device rating, the next higher rating can be used. If the full-load current is over 800 amps and is different from a standard overcurrent device rating, the next lower rating must be used.

If the smallest motor on a circuit has adequate branch-circuit protection, additional loads or motors can be added to the circuit. However, each motor must have overload protection, and it must be ensured that the branch-circuit protective device won't open under the most stressful normal conditions.

Two or more motors (each motor having its own overload protection) or other loads are allowed to be connected to a single circuit in the following cases:

1. The overload devices are factory installed and the branch-circuit, short-circuit, and ground-fault protection are part of the factory assembly or are specified on the equipment.
2. The branch-circuit protective device, motor controller, and overload devices are separate field-installed assemblies, are listed for this use, and are provided with instructions from the manufacturer.
3. All overload devices are marked as suitable for group installation and are marked with a maximum rating for fuses and/or circuit breakers. Each circuit breaker must be of the inverse type and must be listed for group installation.
4. The branch circuit is protected by an inverse time circuit breaker rated for the highest-rated motor and all other loads (including motor loads) connected to the circuit.
5. The branch-circuit fuse or inverse time circuit breaker is not larger than allowed for the overload relay that protects the smallest motor in the group. (See *Section 430-53* of the NEC.)

For *group installation* as described above, taps to single motors don't need branch-circuit protection in any of the following cases:

1. The conductors to the motor have an ampacity equal to or greater than the branch-circuit conductors.
2. The conductors to the motor are no longer than 25 ft, are protected, and have an ampacity at least one-third as great as the branch-circuit conductors.

Adjustable-Speed Drive Systems

The size of branch circuits or feeders to adjustable-speed drive equipment must be based on the rated input current to the equipment. If overload protection is provided by the system controller, no further overload protection is required.

The disconnecting device for adjustable-speed drive systems can be installed in the incoming line; it must be rated at least 115 percent of the conversion unit's input current.

Part-Winding Motors

If separate overload devices are used with standard part-winding motors, each half of the windings must be separately protected at one-half the trip current specified for a conventional motor of the same horsepower rating. Each winding must have separate branch-circuit, short-circuit, and ground-fault protection at no more than one-half the level required for a conventional motor of the same horsepower rating.

A single device (which has the one-half rating) can be used for both windings if it will allow the motor to start.

If a single time-delay fuse device is used for both windings, it can be rated no more than 150 percent of the motor's full-load current.

Torque Motors

The rated current for torque motors must be the locked-rotor current. This nameplate current must be used in determining branch-circuit ampacity, overload, and ground-fault protection.

Adjustable-Voltage AC Motors

The ampacity of switches and branch-circuit, short-circuit, and ground-fault protection for these motors must be based on the full-load current shown on the motor's nameplate. If no nameplate is present, these ratings must be calculated as no less than 150 percent of the values shown in *Tables 430.149* and *430.150* of the NEC.

Motor and Ampacity Ratings

Every motor is considered to be for continuous duty unless the characteristics of the equipment it drives ensure that the motor can't operate with a continuous load.

Except for torque motors and ac adjustable-voltage motors, the current rating of motors (used to determine conductor ampacities, switch ratings, and branch-circuit ratings) must be taken from *Tables 430.147* through *430.150* of the NEC. These values may not be taken from a motor's nameplate rating, except for shaded-pole and permanent split-capacitor fan or blower motors, which are rated according to their nameplates.

Separate overload protection for motors is to be taken from the motor's nameplate rating.

Multispeed motors must have conductors to the line side of the controller rated according to the highest full-load current shown on the motor's nameplate (as long as each winding has its own overload protection, sized according to its own full-load current rating). Conductors between the controller and the motor are to be based on the current for the winding supplied by the various conductors.

Motor Control Circuits

Motor control circuits are circuits that turn the motor on or off, controlling the operation of the motor. A typical control circuit is shown in Fig. 4-5. Note that when the circuit is energized, a set of contacts in parallel with the start switch makes contact, keeping the circuit continuous even when the start switch is no longer depressed.

Note also the overload contacts in the control circuit. If any one of these contacts breaks, it will open the control circuit and cause the operation of the motor to cease.

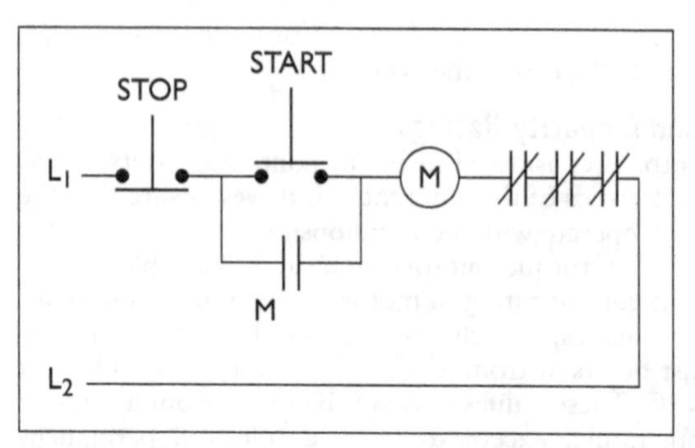

Fig. 4-5 Control circuit.

The rules for motor control circuits are as follows:

Motor control circuits that are *tapped from the load side* of a motor's branch-circuit device and control the motor's operation are not considered branch circuits and can be protected by either a supplementary or a branch-circuit protective device. Control circuits *not tapped this way* are considered signaling circuits and must be protected accordingly. (See *Article 725* of the NEC.)

Motor control conductors as described above must be protected (usually with an in-line fuse) in accordance with Column A of *Table 430.72(B)* of the NEC, except:

1. If they extend no farther than the motor controller enclosure, they can be protected according to Column B of *Table 430.72(B)*.
2. If they extend farther than the motor controller enclosure, they can be protected according to Column C of *Table 430.72(B)*.
3. Control circuit conductors taken from single-phase transformers that have only a 2-wire secondary are considered to be protected by the protection on the

primary side of the transformer. However, the primary protection ampacity should not be more than the ampacity shown in *Table 430.72(B)* multiplied by the secondary-to-primary voltage ratio (secondary voltage divided by primary voltage).

4. When the opening of a control circuit would cause a hazardous situation (as would be the case with a fire pump, for example), the control circuit can be tapped into the motor branch circuit with no further protection.

Control transformers must be protected according to *Article 450* or *Article 725* of the NEC, except:

1. Control transformers that are an integral part of a motor controller and rated less than 50 volt-amperes can be protected by primary protective devices, impedance-limiting devices, or other means.
2. If the primary rating of the transformer is less than 2 amps, an overcurrent device rated at no more than 500 percent of the primary current can be used in the primary circuit.
3. By other approved means.
4. When the opening of a control circuit would cause a hazardous situation (as would be the case with a fire pump, for example), protection can be omitted.

When *damage* to a control circuit would create a hazard, the control circuit must be protected (by raceway or other suitable means) outside of the control enclosure.

When *one side of a motor control circuit is grounded,* the circuit must be arranged so that an accidental ground won't start the motor.

Motor control circuits must be arranged so that they will be shut off from the current supply when the disconnecting device is in the open position.

Motor Controllers

Suitable controllers are required for all motors. The simplest controller is just the branch-circuit protective device, which can be used as a controller for motors of ⅛ horsepower or less that are normally left running and that can't be damaged by overload or failure to start. Another simple "controller" is just a cord-and-plug connection. This can be used for portable motors of ⅓ horsepower or less.

Controllers must have *horsepower ratings* no lower than the horsepower rating of the motor they control, except:

1. Stationary motors of 2 horsepower or less and 300 V or less can use a general-use switch that has an ampere rating at least twice that of the motor it serves. General-use ac snap switches can be used on ac circuits to control a motor rated 2 horsepower or less and 300 V or less and having an ampere rating of no more than 80 percent of the switch rating.
2. A branch-circuit inverse time circuit-breaker rated in amperes only (no horsepower rating) can be used.

Unless a controller also functions as a disconnecting device, it does *not* have to open all conductors to the motor.

If power to a motor is *supplied by a phase converter*, the power must be controlled in such a way that in the event of a power failure, power to the motor is cut off and can't be reconnected until the phase converter is restarted.

Each motor must have its own controller except when a group of motors (600 V or less) uses a single controller rated at no less than the sum of all motors connected to the controller. This applies only in the following cases:

1. If a number of motors drive several parts of a single machine.
2. When a group of motors is protected by one overcurrent device, as specified elsewhere.
3. When a group of motors is located in one room and within sight of the controller.

A controller must be completely capable of stopping and starting the motor and of interrupting its locked-rotor current.

The disconnecting decive must be located within sight of the controller location and within sight of the motor, except in the following situations:

1. If the circuit is more than 600 V, the controller disconnecting device can be out of sight of the controller, as long as the controller has a warning label that states the location of the disconnecting device that is locked in the open position.
2. One disconnecting device can be located next to a group of coordinated controllers on a multimotor continuous process machine.

The disconnecting device for motors 600 V or less must be rated at least 115 percent of the full-load current of the motor being served.

A controller that operates motors of *more than 600 V* must have the control circuit voltage marked on the controller.

Fault-current protection must be provided for each motor operating at over 600 V. (See *Section 430.125(C)* of the NEC.)

All *exposed live parts* must be protected. (See *Sections 430.132* and *430.133* of the NEC, if necessary.)

Tips on Selecting Motors

The first step in selecting a motor for a particular drive is to obtain the following data:

Load. Always try to use a motor near its full load. Motors generally operate at their best power factor and efficiency when fully loaded.

Torque. The starting torque needed by a load must be less than the required starting torque of the proposed motor. Motor torque must never fall below a driven machine's torque needs in going from a standstill to full speed.

The torque requirements of some loads may fluctuate within wide limits. Although average torque may be low, many torque peaks may be well above full load torque. If load torque impulses are repeated frequently, it is usually preferable to use a high-slip motor with a flywheel. But if the load is more or less steady at full load, a more efficient low-slip motor can be used.

Enclosures. The atmospheric conditions that surround a motor determine the type of enclosure that you must use. A motor that is completely enclosed not only costs more than an open motor but runs hotter as well. Totally enclosed motors require a larger frame per horsepower size than open types of motors.

Insulation. The type of insulation used in a motor's construction is determined by the surrounding conditions. Unless stated otherwise, the ambient temperature of a motor is assumed to be 40°C. The temperature of a motor drastically affects its life span. (Insulation tends to break down at higher temperatures.) Each 10°C higher ambient temperature will cut the effective life of a motor in half. Motor temperature ratings are based on the maximum ambient temperature of the motor, measured with an external thermometer.

Varying loads. When the load on a motor varies according to some regular cycle, it is not generally economical to use a motor that matches the peak load. In such cases, the average load can be calculated using the root-mean-square (RMS) method, and a motor (constant-speed motors only) that fits the average load size is chosen.

Fig. 4-6 displays a list of details to be considered for motor installations.

Requirements of the driven machine:

1. Necessary horsepower.
2. Torque requirements.
3. Frequency of stops and starts (also called operating cycle).
4. Speed.
5. Mounting position. (Horizontal, vertical, etc.)
6. Direction of rotation.
7. Ambient temperature.
8. Ambient conditions (water, gas, dust, corrosive elements).

Electrical requirements:

1. Voltage.
2. Phase requirements.
3. Frequency.
4. Starting current.
5. Effect of demand on utility charges.

Fig. 4-6 Motor application considerations.

Characteristics of Squirrel Cage Motors

Squirrel cage induction motors are by far the most commonly used type. Below are some technical details on their construction and classifications.

Squirrel cage motors are classified by the National Electrical Manufacturers Association (NEMA) according to locked-rotor torque, breakdown torque, slip, and starting current. Common types are Classes B, C, and D.

CLASS B, the most common type, has normal starting torque and low starting current. Locked-rotor torque (minimum torque at standstill and full voltage) is not less than 100 percent of full load for 2- and 4-pole motors, 200 hp or less; 40 to 75 percent for larger 2-pole motors; and 50 to 125 percent for larger 4-pole motors.

CLASS C features high starting torque (locked-rotor torque over 200 percent) and low starting current. Breakdown torque is not less than 190 percent of full-load torque. Slip at full load is between 1½ and 3 percent.

CLASS D squirrel cage motors have high slip, high starting torque, and low starting current; they are used on loads with high intermittent peaks. Driven machines usually have a high-inertia flywheel. A no-load motor has little slip; when peak load is applied, motor slip increases. Speed reduction lets driven machines absorb energy from flywheels rather than from power lines.

Full-voltage, across-the-line starting is the most popular method for squirrel cage motors. It is used where the power supply permits and when full-voltage torque and acceleration are not objectionable. Reduction in starting kVA cuts locked-rotor and accelerating torques.

Wound-Rotor Motors

Another popular type of induction motor is the wound-rotor motor. It resembles a dc motor in that both the field and rotor have windings (as opposed to the squirrel cage motor, whose rotor has no windings at all).

The wound-rotor (slip-ring) induction motor's rotor winding connects through slip-rings to an external resistance that is cut in and out by a controller. The advantage of this motor is that the speed of the motor can be easily varied by adding an external resistance to the rotor circuit. Squirrel cage motors can't have their speed varied nearly as easily.

The resistance of the rotor winding affects torque development at any speed. A high-resistance rotor gives high starting torque with low starting current. But low slip at full load, good efficiency, and moderate rotor heating require a low-resistance rotor. This type of motor operates like a squirrel cage motor when all resistance is shorted out.

With the resistance left in, motor speed is decreased below synchronous speed (the speed of the motor's rotating magnetic field).

In addition to high-starting-torque and low-starting-current applications, wound-rotor motors are used in three specific circumstances:

1. With high-inertia loads where high-slip losses would have to be dissipated in the rotor of a squirrel cage motor.
2. Where frequent starting, stopping, and speed control are needed.
3. For continuous operation at reduced speed.

Across-the-line starters are the most common controlling methods for wound-rotor motors. Usually, these starters have a secondary level of speed control with 5 to 7 resistance steps.

Synchronous Motors

Synchronous motors are very similar to wound-rotor motors. They run at a fixed or synchronous speed determined by line frequency and the number of poles in the machine (rpm = 120 × frequency/number of poles). Speed is kept constant by the locking action of an externally excited dc field. Efficiency is 1 to 3 percent higher than that of same-size-and-speed induction or dc motors. Synchronous motors are operated at power factors from 1.0 down to 0.2 leading, for plant power-factor correction. Standard ratings are 1.0 and 0.9 leading power factor. Machines rated down near 0.2 leading are commonly called *synchronous condensers*.

Pure synchronous motors are not self-starting, so in practice they are built with damper or amortisseur windings. With the field coil shorted through the discharge resistor, the damper winding acts like a squirrel cage rotor to bring the motor almost up to synchronous speed. Then the field is applied and the motor pulls into synch, provided that the motor has developed sufficient pull-in torque. Once in synch, the motor keeps constant speed as long as load torque does not exceed maximum (*pull-out*) torque; then the machine drops out of synch. The driven machine is usually started without load. Low-speed motors may be direct-connected.

At constant power, increasing the dc field current in a synchronous motor causes power factor to lead, which is often used for power-factor correction. In some instances, synchronous motors are run with no load or machine connected to them at all—simply to help the facility's power factor.

Decreasing the field current tends to make power factor lag, but either method increases copper losses.

The most common polyphase synchronous motors in general use are:

1. High-speed motors, 500 rpm and up, either general-purpose (500 to 1800 rpm, 200 hp and below) or high-speed units over 200 hp, including most 2-pole motors.
2. Low-speed motors below 500 rpm.
3. Special high-torque motors.

Small, high-speed synchronous motors cost more than comparable induction motors. On the other hand, large, low-speed synchronous motors are generally less expensive than induction motors.

dc Motors

The three basic types of dc motors are shunt motors, series motors, and compound motors. These three motor types are illustrated in Figs. 4-7, 4-8, and 4-9.

Assuming that the normal power source for an installation is ac, the main reason for using dc motors generally lies in the wide ranges of speed control and starting torque. But for constant-speed service, ac motors are generally preferred because they are more rugged and have lower initial cost.

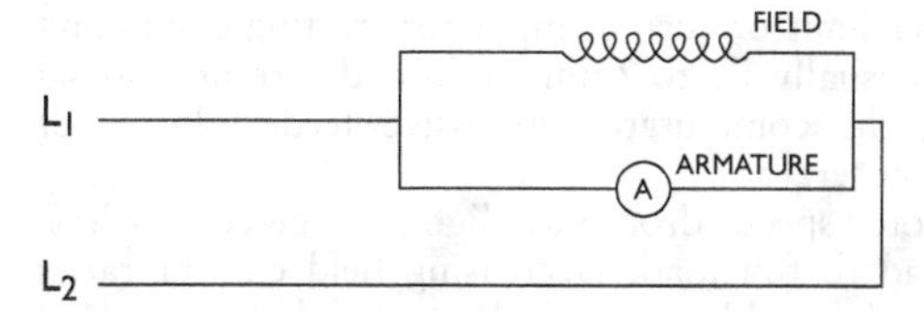

Fig. 4-7 Shunt dc motor.

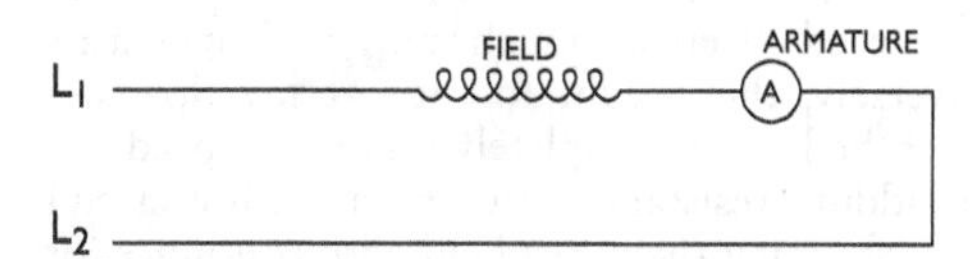

Fig. 4-8 Series dc motor.

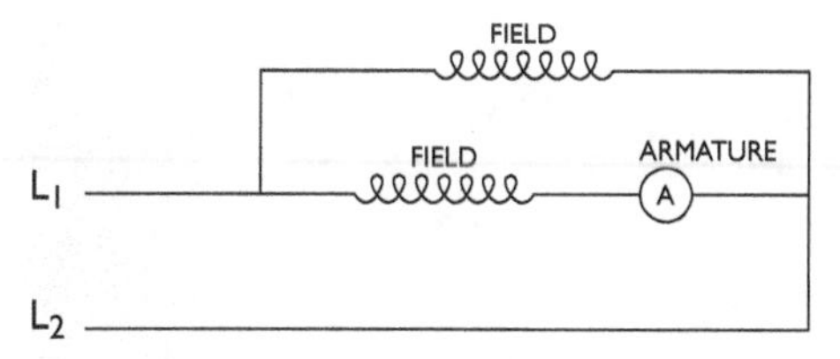

Fig. 4-9 Compound dc motor.

With a shunt dc motor, the torque is proportional to armature current. However, the flux of a series field is affected by the current through it. Compound dc motors (usually the cumulative compound type) lie between the shunt and series motors as to the torque they produce.

The upper limit of current input for starting compound dc motors is usually 1.5 to 2 times full-load current to avoid overheating the commutator, excessive feeder drops, or peaking the generator.

Shunt motor speeds drop only slightly (5 percent or less) from no load to full load. Decreasing field current raises speed; increasing field current reduces speed. But speed is still practically constant for any one field setting. Speed can be controlled by resistance in the armature circuit, although the speed regulation is poor.

Series motor speeds decrease much more with increased load, and conversely, they begin to race at low loads—dangerously so if the load is completely removed. Speed can be reduced by adding resistance into the armature circuit and increased by shunting the series field with resistance, or by short-circuiting series turns.

Compound motors have less constant speed than shunt motors and can be controlled by a shunt-field rheostat.

5. GENERATORS

Utility companies use very large generating units. Generators are also commonly maintained for use in critical or emergency situations. Typical uses would be as backup power systems (UPS systems), for important computers, emergency power for hospitals, and emergency power for lighting in theaters and places of assembly. Generators are also used to provide backup power for other applications. Smaller gasoline-powered generators are the most common types, used either as a portable power source or in more permanent installations where the power requirements are minimal or intermittent.

How Generators Work

The underlying principles governing the operation of generators are the same as those governing the operation of motors. All electric generators operate by using electromagnetic induction, which, simply put, is the interaction between conductors, currents, and magnetic fields. Any time an electrical current passes through a conductor (of which copper wires are the most common type), it causes a magnetic field to form around that conductor. This is one of the absolute laws of physics. Conversely, any time a magnetic field moves through a conductor, it induces (causes to flow) an electrical current in that conductor. This is another absolute and unchangeable law of physics.

The manipulation of these two laws, in effect the combination of magnetic attraction and repulsion, is the basis of the operation of generators. The operation of a generator is such that it turns physical force into an electrical current.

The step-by-step operation of an electric generator follows this pattern: A stationary magnetic field is set up in a generator's windings by causing a current to flow through them. Then, some type of physical force is used to turn the generator's shaft. Coils of wire mounted on this shaft pass through the stationary magnetic field, which induces a current into them. The current is then taken out of the generator (usually by means of a "slip ring" arrangement) and is used for some constructive purpose.

While there are any number of variations and modifications to these basic operations, these are the principles by which all generators function. Depending on the type of generator design, you can increase or decrease power or operate at different voltages.

Installation

Generators are covered in *Article 445* of the National Electrical Code (NEC). Remember, however, that a number of other sections in the Code have specific requirements for generators for specific applications. These sections are frequently those that apply to emergency power applications.

Location

Generators must be of a suitable type for the areas in which they are installed.

Overcurrent Protection

Constant-voltage generators (which includes virtually all now in use) must have overcurrent protection provided by inherent design, circuit breakers, or other means. Alternating-current generator exciters are excepted.

Two-wire dc generators may have only one overcurrent device if it is actuated by the entire current generated, except for current in the shunt field. The shunt field is not to be opened.

Generators that put out *65 volts (V) or less* and are driven by an electric motor are considered protected if the motor driving them will trip its overcurrent protective device when the generator reaches 150 percent of its full-load rated current.

Installation Methods

The ampacity of conductors from the generator terminals to the first overcurrent protective device must be at least 115 percent of the generator's nameplate-rated current. This applies only to phase conductors; neutral conductors can be sized for only the load they will carry. (See *Section 220.22* of the NEC.)

Live parts operating at more than 50 V must be protected. Guards are to be provided where necessary.

Bushings must be used where wires pass through enclosure walls.

Generators and Alternators

Before going too far, we will explain the terms used to describe electrical generation devices.

The terms *generator, alternator,* and *dynamo* have a wide degree of overlap and thus can be pretty confusing. They are often used interchangeably, adding to the confusion. Let's start with "textbook" definitions of these terms before we explain how they are actually used.

> **Generator.** A device that changes mechanical energy into electrical energy.
>
> **Alternator.** A device that changes mechanical energy into electrical energy of the alternating current (ac) type; a generator that produces ac power.
>
> **Dynamo.** A device that changes mechanical energy into electrical energy of the direct current (dc) type; a generator that produces dc power.

These are the "correct" definitions of these terms. In actual practice "in the field," their usage is as follows:

> **Dynamo.** This term is now used only occasionally. This is an old term that was used during the first years of the electrical industry. It is usually used correctly to connote a dc generator, but it may also be used (usually by an older person) to mean any type of electrical generator.

Alternator. This term is usually used correctly. It is sometimes confused with a type of motor control called an "alternator." The motor control "alternator" is used to control two electric motors that are to be used alternately, that is, one after the other.

Generator. This is the catchall term commonly used to describe any type of electrical production device. When this term is used, make sure that you *know* (not just assume you know) the specific type of system that is being referred to.

Sizing a Gasoline Generator

The first consideration in sizing an engine/generator set (the size is measured in watts or VA) is that it must be sufficient for the load it will be called upon to supply. This sizing is complicated by the fact that electric motors require a lot of current when they are started.

A starting current lasts for only a second or so, but it can be as much as four times stronger than the normal running current for small motors. (Very large motors can have much higher starting currents.) This starting surge is an important factor in sizing generators. The starting factor is unimportant when using utility company power, however, because utility company lines have a large capacity that makes the starting currents for common motors too small to notice. (In truth, it can be important, but only if you are talking about motors of several thousand horsepower.)

Engine/generator sets, on the other hand, have their capacity limited by the horsepower of the engine and the inertia of the engine's moving parts. Such generators can handle a surge of short duration, but can't supply a longer-term surge, such as would be demanded by starting a large motor.

If you have a motor that has a running current of 2200 W (about three horsepower), it may seem reasonable that you could supply the necessary current for this motor from a

2500 W (or VA) generator. But if you tried to do so, you would stand a good chance of damaging the motor. When the motor runs, it requires only 2200 W, which is the current rating shown on its nameplate. When it is started, however, it may require 5000 W—far more than the generator is capable of supplying. If the generator is overloaded (as would be the case in this example), the voltage of the output electricity drops; this can often cause motor burnout.

Although it may seem logical that lowering the voltage to a motor would reduce the threat of burnout (less electricity equaling less of a threat), this is not the case. Lowering the voltage to a motor alters the strength of the magnetic field that causes the motor to rotate. Briefly and simply stated: When the voltage is lowered, the effect on the magnetic fields causes a large and damaging current to flow through the motor's windings. In addition to damaging the motor, overcurrent demand on a generator can cause damage to that machine as well.

To properly size the generator for its load, follow these steps:

1. Add the wattages of all lighting, appliances, tools, etc., that will receive power from the generator at the same time. This information can be found on the nameplates of the various pieces of equipment.
2. Determine the wattage of all electric motors that will be fed by the generator.
3. Add the results of steps 1 and 2 to determine your running current requirements.
4. Determine the starting current for each of the motors that will be run off the generator. This starting current averages around three times the running current, or about half of the locked rotor current (shown as *LRA* (Locked Rotor Amps) on most motor nameplates).
5. Determine if more than one motor will ever be started simultaneously.

6. If two or more motors will start simultaneously, use the *extra* starting currents (the difference between their running currents and their starting currents) as your starting currents.
7. If only one motor will ever be started at any one time (or if there is only one motor), use the extra starting current of the largest motor as your starting current.
8. Add 25 percent to both the running and starting wattage totals for future needs.
9. Your generator should have a "rated watts" of at least the running wattage (the extra 25 percent already being added in).
10. Your generator should have a maximum VA of at least the running wattage plus the starting wattage.

Common Wattages

The following is a list of the most common loads and their respective wattages, both running and maximum:

Item	*Running Watts*	*Maximum VA*
100 W bulb	100	100
Radio	150	150
Fan	200	600
Television	400	400
Refrigerator	400	1200
⅓ HP furnace fan	400	1200
Vacuum cleaner	600	1800
⅓ HP sump pump	700	2100
Refrigerator/freezer	800	2400
6-in. circular saw	800	2400
Floodlight	1000	1000

Item	Running Watts	Maximum VA
½-in. drill	1000	3000
Toaster	1200	1200
Coffee maker	1000	1000
Electric skillet	1200	1200
Electric chain saw	1200	3600
½ HP well pump	1400	4200
Single-burner hotplate	1500	1500
10-in. circular saw	2000	6000
Water heater	5000	5000
Electric oven	10,000	10,000

Optional Features

As stated earlier, there are any number of extra features that can be added to gasoline generators. Following is a brief listing of the most common features and the reasons for their use.

Purpose	Feature
Circuit breaker	Protects generator and load from overcurrent situations.
Simultaneous	Allows the operation of a load plus ac and dc power battery charging at the same time.
Panel-mounted outlets	Allow simple access to power receptacles.
Full capacity outlet	Ability to draw the full load from one outlet.
Fully protective cradle	Protects engine and generator in transit.

(continued)

(continued)

Purpose	Feature
Anti-vibration mounting	Reduces vibrations from generator (important when mounting for indoor installations) and helps generator run more smoothly.
Spark arrest mufflers	Eliminates the risk of fire when used near combustible materials; provides quieter operation.
Large gas tank	Longer operating time before refueling.
Low oil shutoff	Prevents engine damage if the oil level drops too low.
Idle control	Reduces engine rpm by half when not operating under load; reduces noise level and fuel consumption; extends engine life.
High motor starting	Special generator windings provide extra surge power for induction and capacitor motor starting.
Electronic ignition	No "points"; reduces maintenance.
Electric start	Turns on by the push of a button.
Industrial duty	Comes with cast iron sleeve; increases engine life.
Suction cooling	Removes heat and fumes from enclosed or system installations.
Wheel kit	Eases transport.
Lifting eye	For larger generators; makes moving unit easier.

Some other features that you should look for are inherent voltage regulation and a continuous-duty rating.

Other Factors

In addition to the factors already discussed, it is critically important that a generator be suited to its environment. If the generator is to be used outdoors, it must have a weatherproof enclosure. It must not be subject to mechanical damage, and if the circuits it feeds run outdoors, it would be wise to install a lightning arrestor. (Even if the outdoor circuits are not likely to be directly struck by lighting, the arrestor remains important because of the voltages that lightning induces into outdoor circuits. Such voltages show up in long lateral runs, such as circuits to well pumps.)

Another factor that must be remembered is altitude. For every 1000 ft of elevation above sea level, a generator's output must be decreased by 3½ (.035) percent.

Generators that turn slowly tend to last longer than ones that turn quickly. Slow-turning generators show less wear in the bearings. This is often a good reason for choosing a slower-turning generator. In addition, fast changes in generator speed can put difficult levels of stress on the generator.

Diesel, Propane, Natural Gas

Most of our discussions about gasoline generators can be applied to diesel, natural gas, and propane generators as well. There are some differences, but they are fairly minor.

Diesel generators tend to be less expensive to operate than gasoline units. They also tend to require less maintenance, in that they avoid conventional ignition systems. The drawback to diesel fuel is that it is sometimes harder to procure than gasoline.

Gas-driven generators (propane or natural gas) are also less expensive to operate than are gasoline units. Like diesel generators, gas-driven generators are more expensive than gasoline-driven units, although this cost difference is not terribly great. The extra cost of a propane generator, compared to a gasoline unit, is only in the $200 range for a 4-kW or 5-kW unit.

Procuring the required propane (or natural gas) can be a problem—or it can be extremely convenient. If the home you are installing the generator in uses natural gas or propane for cooking and/or heating, just tap into the gas piping. If there is no gas supply at the house, getting a consistent supply can be difficult.

Synchronous Generators

The synchronous generator is most commonly used for mid-sized independent generating systems, by virtue of its reliability and flexibility. Synchronous generators have a rotating field. That is, the armature windings (at the center of the generator) are held stationary, and the field windings (at the perimeter of the generator) are rotated. This is the opposite of the electric motors you are most familiar with, where the armature in the middle of the motor moves and the field windings are stationary.

Thus, the electricity produced by these units is taken directly from the stationary armature rather than from the moving field. This is often preferred, because taking electricity from an alternator requires the use of slip rings and brushes (which require maintenance once every five years). It is easier to service a rotating field winding than it is to service a stationary armature.

The variables of frequency, voltage, and current put out by synchronous generators are:

> **Frequency.** Determined by the speed of the field's rotation.
>
> **Voltage.** Determined by the amount of current flowing through the field windings.
>
> **Current.** Determined by the amount of torque that is necessary to move the field windings.

Note that the current that flows through the synchronous generator's field windings is direct current (dc). Using alternating current (ac) in these windings would produce wildly fluctuating frequencies and voltages and would guarantee a speedy failure of the unit. The synchronous generator produces alternating currents when the magnetic fields produced by the dc currents in the field windings pass in and out of the armature windings. When the magnetic fields move into the armature windings they produce current in one direction, and when they move out of the armature they produce current in the other direction.

Notice that the amount of current that is produced by this type of generator is directly related to the amount of torque necessary to turn the field. In other words, when the load you are serving requires more current, the generator shaft will become more difficult to turn. And conversely, when your load needs less current, the field windings will become easier to turn. This effect is caused by the interaction of magnetic fields in the generator.

Generator Exciters

Practically all modern synchronous generators use a brushless *exciter,* which creates the initial current in the generator's windings. The exciter is a small ac generator on the main shaft. The ac voltage generated is "rectified" (turned into direct current) by a 3-phase rotating rectifier assembly, also on the shaft. The dc voltage thus obtained is applied to the main generator field, which is also on the main shaft. A voltage regulator is provided to control the exciter field current, and in this manner the field voltage can be precisely controlled, resulting in a stable, well-controlled generator output voltage.

Because of recent problems with "nonlinear" (fluctuating) loads that cause unwanted harmonic currents in the power system, some generator manufacturers are providing an additional generator furnished with a permanent magnet exciter (PME) to provide field magnetism to the brushless exciter. When selecting a synchronous generator, investigate the need for or availability of the permanent magnet exciter.

Voltage Regulation

The output voltage of a synchronous generator is controlled by the excitation in the field windings. To control this, the generator's voltage regulator measures the output voltage, compares it to a standard reference voltage obtained from a zener diode that continuously samples output voltage, and adjusts the excitation current up or down as needed to maintain the output voltage at its rated value. If the load varies, the excitation is continuously adjusted to keep the voltage constant.

Frequency and Frequency Regulation

The frequency of the ac current output is dependent on two factors: the number of magnetic poles built into the machine and the speed of rotation (rpm). Frequency can be calculated as follows:

$$\text{Frequency} = \frac{\text{rpm} \times \text{Number of poles}}{120}$$

As a result, a 2-pole generator must rotate at 1800 rpm to provide 60 Hz, and an 8-pole generator must operate at 900 rpm to provided 60 Hz. To obtain a 50-Hz output, the generator speeds must be slightly slower, as calculations would show.

Since frequency is normally a constant (60 Hz or 50 Hz), control of the generator speed is essential. This is accomplished by providing precise rpm control of the prime mover, which is done by an engine speed control device called a *governor.*

Connections

Synchronous generator connections can vary widely. The machine may be designed for single- or three-phase operation, single or dual voltage, in a wye or delta configuration, etc. Most modern synchronous generators are connected in a wye and many provide for dual voltages.

Lead Identification

In recent years, two standard systems have been used to identify generator leads that are brought out to supply loads: National Electrical Manufacturers Association (NEMA) and International Electrotechnical Commission (IEC). Generators built in the United States adhere to markings in NEMA Standard MGE. These leads are typically identified with a "T" and a numeral. Figure 5-1 shows NEMA markings for a typical generator along with IEC markings that use a U, V, and W with numerals for identification.

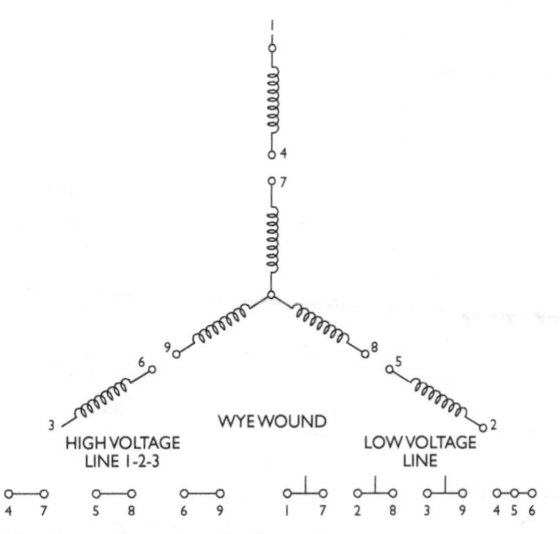

Fig. 5-1 Generator lead markings.

Both NEMA and IEC standards specify that field leads be identified with an "F," with F1 being positive and F2 being negative. These leads may also be marked with a (+) or POS and (–) or NEG.

Specifications

When specifying, sizing, or selecting a generator, many factors must be considered. As previously mentioned, factors that relate to the prime mover must receive equal attention simultaneously. The following are important generator considerations.

- Type of generator
- rpm, frequency
- Location, enclosures
- Kilowatt rating, efficiency
- Number of phases, power factor
- Controls, related switch gear
- Transfer switching
- Duty, starting conditions

Be sure to contact your supplier if the generator will have to supply a large motor load. Some motors draw very large starting currents that will place heavier demands on the generator than it can handle.

Nonlinear Loads

If the facility loads include nonlinear loads, such as computer power supplies, variable-frequency drives, electronic ballasts, or other similar electronic equipment (particularly those furnished with "switch-mode power supplies"), it is essential to advise the generator supplier of this so that proper steps can be taken to avoid equipment overheating or other problems caused by harmonics. Some generator manufacturers recommend low-impedance generators and have developed winding design techniques to reduce the effects of the harmonic currents generated. In some instances, the generator may have to be derated and the neutral conductor size increased in order to safely supply complex nonlinear loads.

The Induction Generator

Induction generators are most commonly used when power is to be sold back to a utility company.

Construction of an induction generator is essentially the same as an induction motor with a squirrel cage rotor and wound stator. When this machine is driven above its designed synchronous speed, it becomes a generator. (At less than synchronous speed, it functions as a motor.) Normally, if a rotating machine is to operate as a generator, certain design changes can be incorporated to enable it to operate more effectively and efficiently.

Because the induction generator does not have an exciter, it must operate in parallel with the utility. This outside power source provides the reactive power for generator operation. Also, its frequency is automatically locked in with the utility.

Induction generators are a popular choice for use in cogeneration systems where they will operate in parallel with the utility. They offer certain advantages over synchronous generators. Voltage and frequency are controlled by the utility, so voltage and frequency regulators are not required. The induction generator's construction offers high reliability and little maintenance, and minimal protective relays and controls are required. Its major disadvantage is that it normally can't operate alone as a standby/emergency generator in many applications.

6. MECHANICAL POWER TRANSMISSION

Mechanical power transmission is especially important in an industrial setting. With so much mechanical power (as differentiated from to electrical power) being used for the various processes, its transmission is an important consideration. Mechanical power transmission becomes the responsibility of the electrician when it relates to the equipment he or she installs.

The first concern in mechanical power transmission is that all of the equipment should be firmly and permanently mounted to a solid base. All methods of power transmission covered in this chapter assume that the equipment is properly mounted.

The most common method for the transmission of rotary mechanical power between adjacent shafts is via belts, although chains and gears are also used.

This chapter will cover the use of V belts, general power transmission concepts, and power transmission couplings.

Pitch

Pitch is a word commonly used in connection with machinery and mechanical operations. Pitch is *the distance from one point to a corresponding point.*

Pitch Diameter and Pitch Circle

The pitch diameter specifies the distance across the center of the pitch circle. Pitch diameter dimensions are specific values even though the pitch circle is imaginary. Rotary power transmission calculations are based on the concept of circles or cylinders in contact. These circles are called pitch circles.

As the shafts rotate, the surfaces of the pitch circles travel equal distances at equal speeds (assuming no slippage). Shafts then rotate at speeds proportional to the circumference of the pitch circles and therefore proportional to the pitch diameters.

This concept of pitch circles in contact applies to belt and chain drives, although the pitch circles are actually separated. This is true because the belt or chain is in effect an extension of the pitch circle surface.

As rotation occurs, the pitch circle surfaces will travel the same distance at the same surface speed.

Because the circumference of a 4-in. pitch circle is double that of a 2-in. pitch circle, its rotation will be one-half as much. The rotation speed of the 4-in. pitch circle will be one-half that of the 2-in. pitch circle.

Calculations

Rotational speed and pitch diameter calculations for belts, chains, and gears are based on the concept of pitch circles in contact. The relationship that results from this concept may be stated as follows:

Shaft speeds are inversely proportional to pitch diameters.

In terms of rotational speeds and pitch diameters, this relationship may be expressed in equation form as follows:

$$\frac{\text{Driver rotational speed}}{\text{Driven rotational speed}} = \frac{\text{Driven pitch diameter}}{\text{Driver pitch diameter}}$$

To simplify the use of the equation, letters and numbers instead of words are used to represent the terms.

$S1$ for Driver rotational speed

$S2$ for Driven rotational speed

$P1$ for Driver pitch diameter

$P2$ for Driven pitch diameter

The basic equation then becomes:

$$\frac{S1}{S2} = \frac{P2}{P1}$$

The equation may be arranged into the following forms, one for each of the four values. To find an unknown value, the known values are substituted in the appropriate equation.

$$P1 = \frac{S2 \times P2}{S1} \quad P2 = \frac{S1 \times P1}{S2}$$

$$S1 = \frac{P2 \times S2}{P1} \quad S2 = \frac{P1 \times S1}{P2}$$

The above equations may also be stated as rules. The following are other convenient ways to calculate known shaft speeds and pitch diameters:

To find *Driving Shaft Speed,* multiply the driving pitch diameter by the speed of the driven shaft and divide by the driving pitch diameter.

To find *Driven Shaft Speed,* multiply the driving pitch diameter by the speed of the driving shaft and divide by the speed of the driven pitch diameter.

To find *Driven Pitch Diameter,* multiply the driven pitch diameter by the speed of the driven shaft and divide by the speed of the driving shaft.

V Belts

The V belt has a tapered cross-sectional shape that causes it to wedge firmly into the sheave groove under a load. Its driving action takes place through frictional contact between the sides of the belt and the sheave groove surfaces. While the cross-sectional shape varies slightly with make, type, and size, the included angle of most V belts is about 42°. There are three general classifications of V belts: Fractional Horsepower, Standard Multiple, and Wedge.

Fractional Horsepower

Fractional horsepower belts are designed for intermittent and relatively light loads. They are used principally as single belts on fractional horsepower drives. These belts are manufactured in four standard cross-sectional sizes as illustrated in Fig. 6-1.

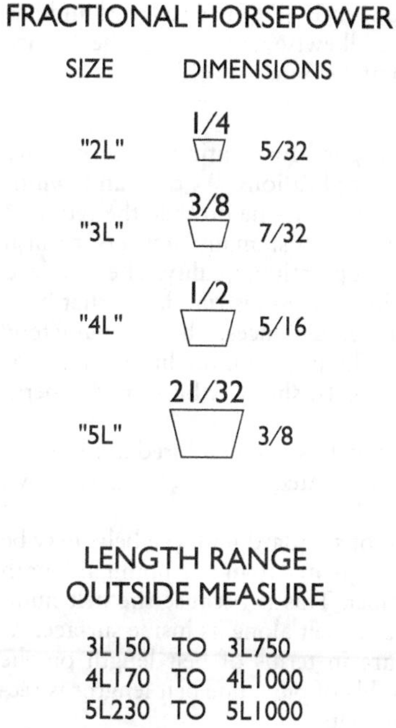

FRACTIONAL HORSEPOWER

SIZE	DIMENSIONS	
"2L"	1/4	5/32
"3L"	3/8	7/32
"4L"	1/2	5/16
"5L"	21/32	3/8

LENGTH RANGE
OUTSIDE MEASURE

3L150 TO 3L750
4L170 TO 4L1000
5L230 TO 5L1000

Fig. 6-1 V belt sizing.

Standard belt lengths vary by one-inch increments between a minimum length of 10 in. and a maximum length of 100 in.

The numbering system used indicates the cross-sectional size and the nominal outside length. The last digit of the belt number indicates tenths of an inch. Because the belt number indicates length along the outside surface, belts are slightly shorter along the pitch line than the nominal size number indicates.

For example, a belt number 3L 470 is the 3L size and has an outside length of 47 in. Likewise, a 4L 425 is the 4L size and has an outside length of 42½ in.

Standard Multiple

Standard multiple belts designed for continuous service are usually used for industrial applications. As the name *multiple belts* indicates, more belts than one provide the required power transmission capacity. Most manufacturers furnish two grades: a standard and a premium quality. The standard belt is suitable for the majority of industrial drives that have normal loads, speeds, center distances, sheave diameters, and operating conditions. The premium quality is made for drives subjected to severe loads, shock, vibration, temperatures, etc.

The standard multiple V belt is manufactured in five standard cross-sectional sizes designated: A, B, C, D, E, as shown in Fig. 6-2.

The actual pitch length of standard multiple belts may be from one to several inches greater than the nominal length indicated by the belt number. This is because the belt numbers indicate the length of the belt along its inside surface. As belt length calculations are in terms of belt length on the pitch line, reference to a table of pitch line belt lengths is recommended when selecting belts.

Wedge Belt

The wedge belt is an improved design V belt which makes possible a reduction in size, weight, and cost of V belt drives. Utilizing improved materials, these multiple belts have a smaller cross-section per horsepower and use smaller diameter sheaves at shorter center distances than is possible with standard multiple belts. Because of heavy-duty construction, only three cross-sectional belt sizes are used to cover the duty range of the five sizes of standard multiple belts. The dimensions of the three standard wedge belt cross-sectional sizes—3V, 5V, 8V—are shown in Fig. 6-3.

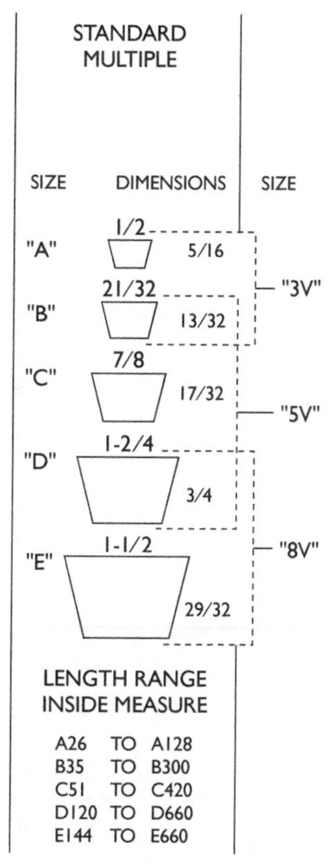

Fig. 6-2 Multiple V belt sizes.

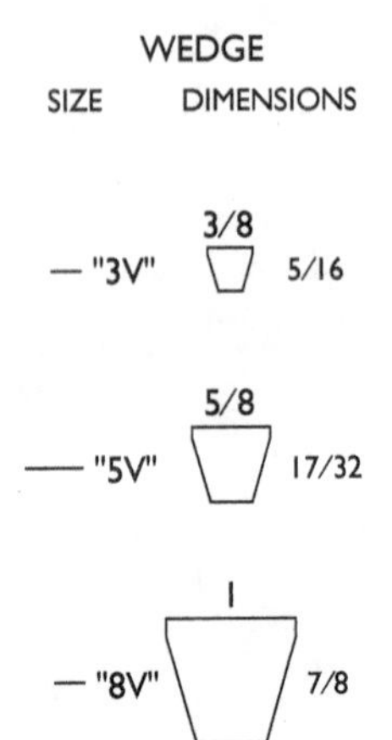

LENGTH RANGE
PITCH LINE MEASURE

3V250 TO 3V1650
5V500 TO 5V3550
8V1000 TO 8V5000

Fig. 6-3 Dimensions of the three standard wedge belt sizes.

The wedge belt number indicates the number of ⅛ in. of top width of the belt. As shown in Fig. 6-4, the 3V belt has a top width of ⅜ in., the 5V a width of ⅝ in., and the 8V a full 1 in. of top width.

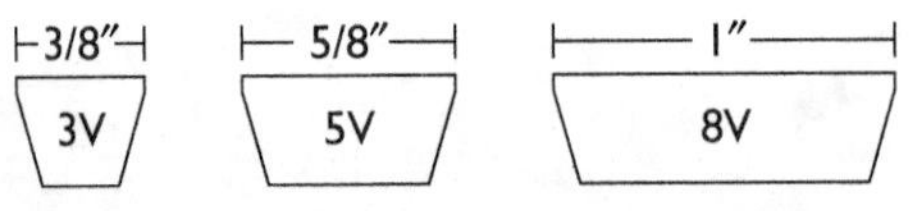

Fig. 6-4 Belt width.

The belt length indicated by the number is the effective pitch line length of the belt. Since belts are identified in terms of pitch line lengths, belt numbers can be used directly when choosing wedge belts.

Belt Matching

Satisfactory operation of multiple belt drives requires that each belt carry its share of the load. To accomplish this, all belts in a drive must be essentially of equal length. Because it is not economically practical to manufacture belts to an exact length, most manufacturers follow a practice of code marking (see Table 6.1).

Each belt is measured under specific tension and is marked with a code number to indicate its variation from nominal length. The number 50 is commonly used as the code number to indicate a belt within tolerance of its nominal length. For each 1⁄10 of an inch over nominal length, the number 50 is increased by 1. For each 1⁄10 of an inch under nominal length, 1 is subtracted from the number 50. Most manufacturers' codes are marked as shown in Fig. 6-5.

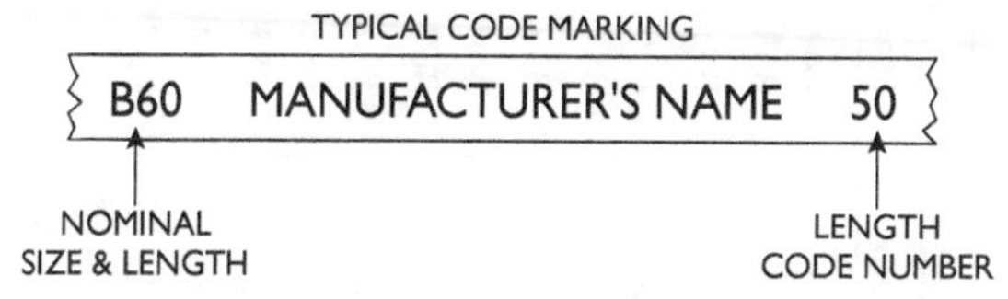

Fig. 6-5 Code markings.

For example, if the 60-in. B section belt shown is manufactured 3⁄10 of an inch longer, it will be code marked 53 rather than the 50 shown. If it is made 3⁄10 of an inch shorter, it will be code marked 47. While both of these belts have the belt number B60, they cannot be used satisfactorily in a set because of the difference in their actual length.

Table 6.1 Belt Code Markings

A Belts			B Belts			C Belts		
Belt Number Standard	*Pitch Length*	*Outside Length*	*Belt Number Standard*	*Pitch Length*	*Outside Length*	*Belt Number Standard*	*Pitch Length*	*Outside Length*
A26	27.3	28.0	B35	36.8	38.0	C51	53.9	55.0
A31	32.3	33.0	B38	39.8	41.0	C60	62.9	64.0
A35	36.3	37.0	B42	43.8	45.0	C68	70.9	72.0
A38	39.3	40.0	B46	47.8	49.0	C75	77.9	79.0
A42	43.3	44.0	B51	52.8	54.0	C81	83.9	85.0
A46	47.3	48.0	B55	56.8	58.0	C85	87.9	89.0
A51	52.3	53.0	B60	61.8	63.0	C90	92.9	94.0
A55	56.3	57.0	B68	69.8	71.0	C96	98.9	100.0
A60	61.3	62.0	B75	76.8	78.0	C105	107.9	109.0
A68	69.3	70.0	B81	82.8	84.0	C112	114.9	116.0
A75	76.3	77.0	B85	86.8	88.0	C120	122.9	124.0
A80	81.3	82.0	B90	91.8	93.0	C128	130.9	132.0
A85	86.3	87.0	B97	98.8	100.0	C136	138.9	140.0

A Belts			B Belts			C Belts		
Belt Number Standard	Pitch Length	Outside Length	Belt Number Standard	Pitch Length	Outside Length	Belt Number Standard	Pitch Length	Outside Length
A90	91.3	92.0	B105	106.8	108.0	C144	146.9	148.0
A96	97.3	98.0	B112	113.8	115.0	C158	160.9	162.0
A105	106.3	107.0	B120	121.8	123.0	C162	164.9	166.0
A112	113.3	114.0	B128	129.8	131.0	C173	175.9	177.0
A120	121.3	122.0	B136	137.8	139.0	C180	182.9	184.0
A128	129.3	130.0	B144	145.8	147.0	C195	197.9	199.0
			B158	159.8	161.0	C210	212.9	214.0
			B173	174.8	176.0	C240	240.9	242.0
			B180	181.8	183.0	C270	270.9	272.0
			B195	196.8	198.0	C300	300.9	302.0
			B210	211.8	213.0	C360	360.9	362.0
			B240	240.3	241.5	C390	390.9	392.0
			B270	270.3	271.5	C420	420.9	422.0
			B300	300.3	301.5			

(continued)

Table 6.1 (continued)

D Belts			3V Belts		5V Belts		8V Belts	
Belt Number Standard	**Pitch Length**	**Outside Length**	**Belt Number**	**Belt Length**	**Belt Number**	**Belt Length**	**Belt Number**	**Belt Length**
			3V250	25.0	5V500	50.0	8V1000	100.0
D120	123.3	125.0	3V265	26.5	5V530	53.0	8V1060	106.0
D128	131.3	133.0	3V280	28.0	5V560	56.0	8V1120	112.0
D144	147.3	149.0	3V300	30.0	5V600	60.0	8V1180	118.0
D158	161.3	163.0	3V315	31.5	5V630	63.0	8V1250	125.0
D162	165.3	167.0	3V335	33.5	5V670	67.0	8V1320	132.0
D173	176.3	178.0	3V355	35.5	5V710	71.0	8V1400	140.0
D180	183.3	185.0	3V375	37.5	5V750	75.0	8V1500	150.0
D195	198.3	200.0	3V400	40.0	5V800	80.0	8V1600	160.0

It is possible for the length of belts to change slightly during storage. Under good conditions, however, these changes won't exceed measuring tolerances. Belts may be combined by matching code numbers. Ideally, sets should be made up of belts having the same code numbers; however, the resiliency of the belts allows some length variation.

Alignment

The life of a V belt is dependent on first, the quality of materials and second, on installation and maintenance. One of the most important installation factors influencing operating life is belt alignment. In fact, excessive misalignment is probably the most frequent cause of shortened belt life.

While V belts, because of their inherent flexibility, can accommodate themselves to a degree of misalignment not tolerated by other types of power transmission, they still must be held within reasonable limits. Maximum life can be attained only with true alignment, and misalignment greater than 1⁄16 in. for each 12 in. of center distance will result in excessive wear.

Misalignment of belt drives results from shafts being out of angular or parallel alignment or from the sheave grooves being out of axial alignment. These three types of misalignment are illustrated in Fig. 6-6.

Because the shafts of most V belt drives are in a horizontal plane, angular shaft alignment is easily obtained by leveling the shafts. In those cases where shafts are not horizontal, a careful check must be made to ensure that the angle of inclination of both shafts is the same.

A check for parallel-shaft and axial-groove alignment of most drives can be done simultaneously if the shafts and sheaves are true.

The most satisfactory method of checking parallel-shaft and axial-groove alignment is with a straightedge. This method is illustrated in Fig. 6-7 with arrows indicating the four check points. When sheaves are properly aligned, no light should be visible at these four points.

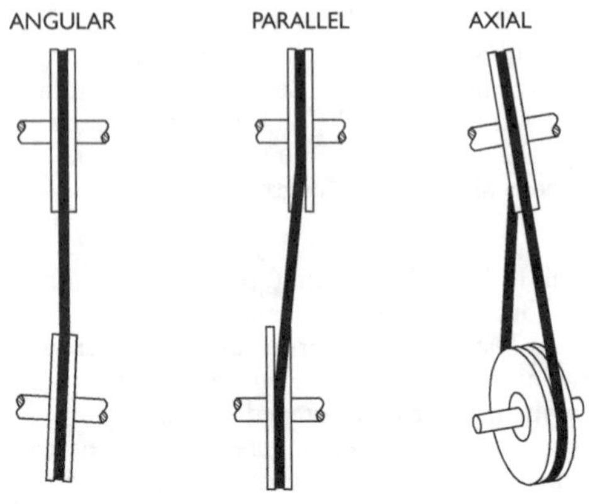

Fig. 6-6 Types of misalignments.

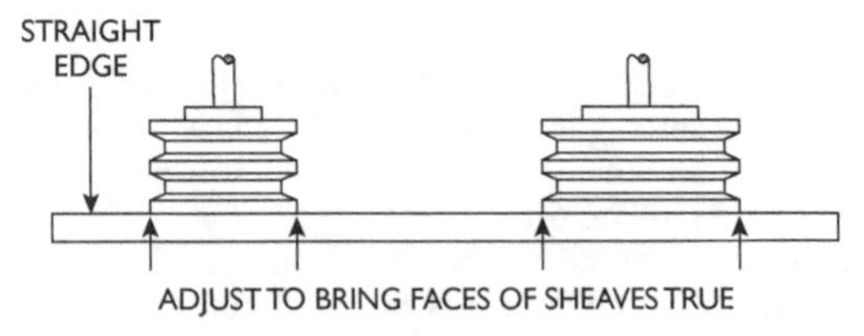

Fig. 6-7 Checking belt alignments.

Installation

V belts should never be "run on" to sheaves. Doing so places excessive stress on the cords, usually straining or breaking some of them. A belt damaged in this manner will flop under a load and turn over in the sheave groove. The proper installation method is to loosen the adjustable mount, reduce the center distance, and slip the belts loosely into the sheave grooves.

The following six general rules should be followed when installing V belts:

1. Reduce centers so belts can be slipped on sheaves.
2. Have all belts slack on the same side (top of drive).
3. Tighten belts to approximately correct tension.
4. Start unit and allow belts to seat in grooves.
5. Stop; re-tighten to correct tension.
6. Recheck belt tension after 24 to 48 hours of operation.

Tension

Belt tension is a vital factor in operating efficiency and service life. Too low a tension results in slippage and rapid wear of both belts and sheave grooves. Too high a tension stresses the belts excessively and unnecessarily increases bearing loads.

The tensioning of fractional horsepower and standard multiple belts may be done satisfactorily by tightening until the proper feel is attained. With proper feel the belt has a live springy action when struck with the hand. If there is insufficient tension the belt will feel loose or dead when struck. Too much tension will cause the belts to feel taut, as there will be no give to them.

Couplings

Power transmission couplings are the usual means of connecting coaxial shafts so that one can drive the other. For example, they are used to connect an electric motor to a pump shaft, or to the input shaft of a tear reducer, or to connect two pieces of shafting to obtain a long length, as with line shafting. Power transmission couplings for such shaft connections are manufactured in a great variety of types, styles, and sizes. They may, however, be divided into two general groups or classifications: rigid (solid) couplings and flexible couplings.

Solid Couplings

Solid couplings, as the name indicates, connect shaft ends together rigidly, making the shafts so connected into a single continuous unit. They provide a fixed union that is equivalent to a shaft extension. They should only be used when true alignment and a solid or rigid coupling are required, as with line shafting, or where provision must be made to allow parting of a rigid shaft. A feature of rigid couplings is that they are self-supporting and automatically align the shafts to which they are attached when the coupling halves on the shaft ends are connected.

There are two basic rules that should be followed to obtain satisfactory service from rigid couplings:

1. A force must be used in assembly of the coupling halves to the shaft ends.
2. After assembly, a runout check of all surfaces of the coupling must be made, and any surface found to be running out must be machined true.

Checking of surfaces is especially necessary if the coupling halves are assembled by driving or bumping rather than by pressing. Rigid couplings should *not* be used to connect shafts of independent machine units that must be aligned at assembly.

Couplings

The transmission of mechanical power often requires the connection of two independently supported coaxial shafts so one can drive the other. Prime movers (usually internal combustion engines or electric motors) connected to pumps, variable-speed drives, and the like are typical examples. For these applications, a flexible coupling is used because perfect alignment of independently supported coaxial shaft ends is practically impossible. In addition to the impracticality of perfect alignment, there is always wear and damage to the connected components and their shaft bearings as well as the possibility of movement due to temperature changes and other external forces.

In addition to enabling coaxial shafts to operate with a slight misalignment, flexible couplings allow some axial movement, or end float, and may allow torsional movement as well. Another benefit that may result when flexible couplings have nonmetallic connecting elements is electrical insulation of connected shafts. This is especially important when being applied to generators that are feeding power to sensitive equipment. In these cases, small transient voltages have been known to travel through metal shafts and into motor or generator windings.

Flexible couplings are designed to compensate for the following conditions:

1. Angular Misalignment
2. Parallel Misalignment
3. Axial Movement (End Float)
4. Torsional Movement

Illustrated in Fig. 6-8 are the four types of flexibility that may be provided by flexible couplings.

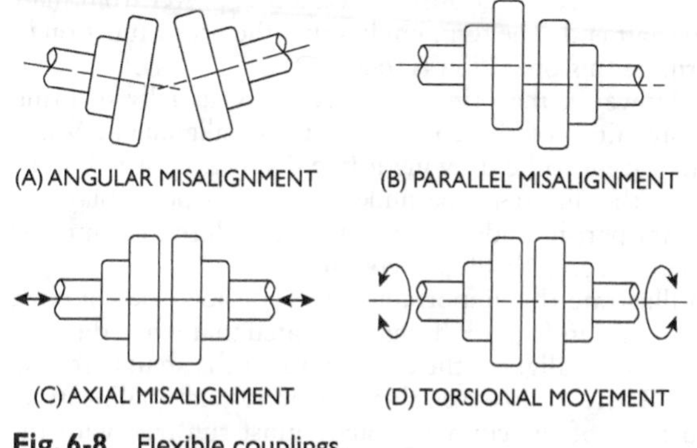

Fig. 6-8 Flexible couplings.

Flexible couplings are intended to compensate only for the slight unavoidable minor misalignment that is inherent in the design of machine components and in the practices followed when aligning coaxial shafts of connected units. If the application is one where misalignment must exist, universal joints, some style of flexible shafting, or special couplings designed for offset operation may be necessary. Flexible couplings should not be used in attempting to compensate for deliberate misalignment of connected shaft ends.

Most flexible couplings are made up of three basic parts: two hubs that attach to the shaft ends to be connected and a flexing member or element that transmits power from one hub to the other. There are a variety of flexible coupling designs, all having characteristics and features to meet specific needs. Most flexible couplings fall into one or more groupings.

Coupling Alignment

The designation "coupling alignment" is the accepted term to describe the operation of bringing coaxial shaft ends into alignment. The most common situation is where a coupling is used to connect and transmit mechanical power from shaft end to shaft end. The term implies that the prime function in performing this operation is to align the surfaces of the coupling. However, the principal objective is actually to bring the center lines of the coaxial shaft into alignment. While this may seem to be drawing a fine distinction, it is important that the difference be understood. This point plays an important part in understanding the procedure that must be followed to correctly align a coupling.

To illustrate this important point, consider the coupling halves shown in Fig. 6-9. It may be stated that when the coupling faces are aligned, the center lines of the shafts are also aligned. There is, however, this very important qualification: the surfaces of the coupling halves must run true with the center lines of the shafts.

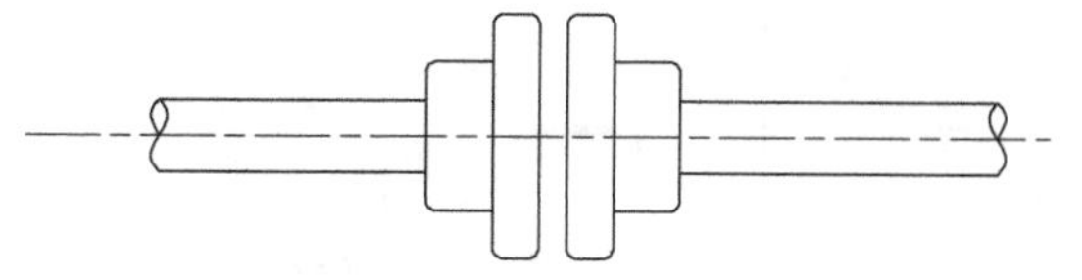

Fig. 6-9 Alignment.

If the surfaces don't run true with the shaft center lines, alignment of the untrue surfaces will result in misalignment of the shaft center lines. Well-aligned running surfaces, therefore, are a basic requirement if the alignment procedures that follow are to be successful.

Shafts may be misaligned in two ways. They may be at an angle rather than in a straight line (angular misalignment), or they may be offset (parallel misalignment). These misaligned shafts are illustrated in Fig. 6-10.

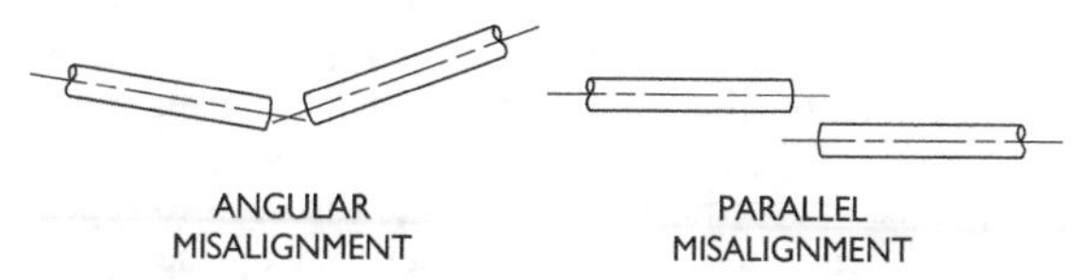

Fig. 6-10 Misalignment.

Angular misalignment may be a tilt up, down, or to the side of one shaft in respect to the other. Parallel misalignment may be one shaft high, low, or to the side of the other. A practical method of correcting this misalignment is to align the center lines in two planes at right angles. The practice commonly followed, because it is convenient, is to check and adjust in the vertical and horizontal planes. The vertical and horizontal planes are illustrated in Fig. 6-11.

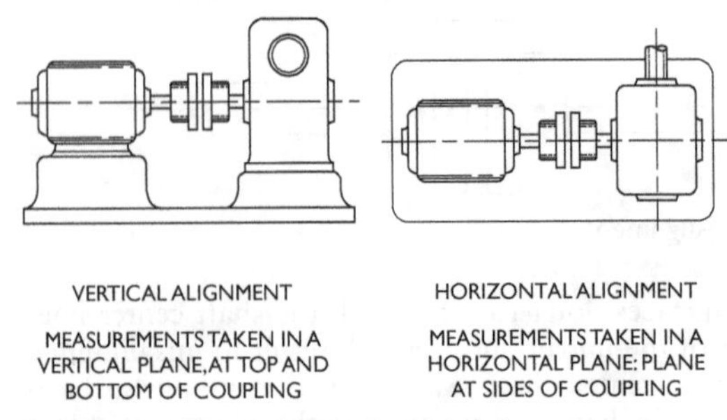

Fig. 6-11 Horizontal and vertical alignment.

To align center lines of coaxial shafts in two planes, vertical and horizontal, requires an angular and parallel alignment in the vertical plane and the same in the horizontal plane. It is a mistake to perform these operations in random order, making adjustments by trial and error. This practice results in a time-consuming series of operations, with some of the adjustments changing settings made during prior adjustments.

An organized procedure can eliminate the need to repeat operations. When a definite order is established, one operation is completed before another is started. In many cases, when the correct order is followed, it is possible to correctly align a coupling by going once through the four operations. When extreme accuracy is required the operations may be repeated, but always in the correct order. As adjustments are made by the insertion of shims at low support points, an important preparatory step is to check the footings of the units. If there is any rocking motion, the open point must be eliminated by shimming so that all support points rest solidly on the base plate.

The standard thickness gauge, also called a feeler gauge, is a compact assembly of high-quality heat-treated steel leaves of various thickness. This is the measuring instrument commonly used to determine the dimension of an opening or gap. The leaves are inserted, singly or in combination, until a leaf or combination is found that fits snugly. Then the dimension is ascertained by the figure marked on the leaf surface or, if several are used, by totaling the surface figures.

Another precision tool, not as widely known or used but ideally suited for coupling alignment, is the taper gauge.

The principal advantage of the taper gauge, sometimes called a gap gauge, for coupling alignment is that it is direct reading, not requiring trial-and-error feeling to determine a measurement. The tip end is inserted into an opening or gap, and the opening size is read on the graduated face. The tool is graduated in one thousands of an inch on one side and millimeters on the other.

Two methods of coupling alignment are widely used: the straightedge-feeler gauge method and the indicator method. In both methods, four alignment operations are performed in a specific order. Only when performed in the correct order can adjustments be made at each step without disturbing prior settings. The four steps in the order of performance are:

1. **Vertical Face Alignment.** The first adjustment is made to correct angular misalignment in the vertical plane. This is accomplished by tipping the unit as required. The gap at the top and bottom of the coupling is measured, and adjustment is made to bring these faces true.

2. **Vertical Height Alignment.** This adjustment corrects parallel misalignment in the vertical plane. The unit is raised without changing its angular position. Height difference from base to center line is determined by measuring on the OD of the coupling at top and/or bottom.

3. **Horizontal Face Alignment.** When the units are in alignment vertically, shimming is complete. The horizontal alignment operations may then be done simultaneously.
4. **OD Alignment.** The unit is moved as required to align the faces and ODs at the sides of the coupling.

Vertical Face Alignment

The initial alignment step can be a time-consuming operation when done in an unorganized manner. Correction of angular misalignment requires tilting one of the units into correct position using shims. Selection of shim thickness is commonly done by trial and error.

The tilt required, or the angle of change at the base, is the same as the angle of misalignment at the coupling faces. Because of this angular relationship, the shim thickness is proportional to the misalignment. For example, in Fig. 6.12 the misalignment at the coupling faces is 0.006 in. in 5 in.; therefore each 5 in. of base must be tilted 0.006 in. to correct misalignment. As the base length is twice 5 in., the shim thickness must be twice 0.006 in., or 0.012 in.

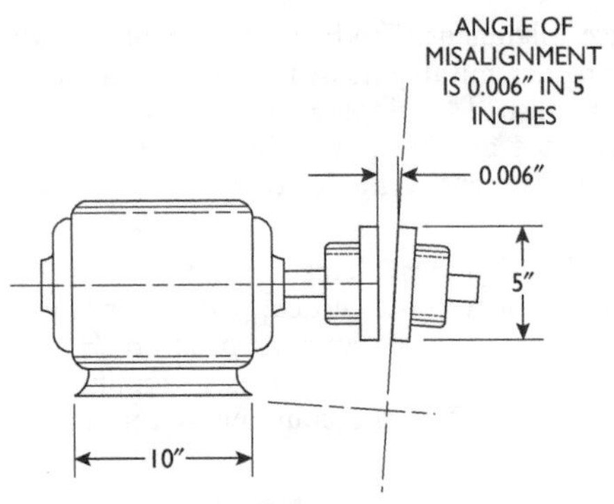

Fig. 6-12 Misalignment.

The general rule for shim thickness is as follows: Shim thickness is as many times greater than the misalignment as the base length is greater than the coupling diameter.

In preparing to align a coupling it must be determined which unit is to be adjusted—the driver or the driven. Common practice is to position, level, and secure the driven unit at the required elevation. Then, adjust the driver to align with it. Connections to the driven unit, such as pipe connections to a pump or output shaft connections to a reducer, should be completed prior to proceeding with coupling alignment. The driven unit should be set with its shaft center line slightly higher than the driver to allow for alignment shims.

Practically all flexible couplings on drives operating at average speeds will perform satisfactorily when misaligned as much as .005 in. Some will tolerate much greater misalignment. Alignment well within .005 in. is easily and quickly attainable using a straightedge and a feeler gauge when the following correct methods are followed.

1. **Vertical Face Alignment.** Using the feeler gauge, measure the width of gap at top and bottom between the coupling faces. Using the difference between the two measurements, determine the shim thickness required to correct alignment. (It will be as many times greater than the misalignment as the driver base length is greater than the coupling diameter.) Shim under the low end of the driver to tilt into alignment with the driven unit.

2. **Vertical Height Alignment.** Using a straightedge and feeler gauge, measure the height difference between driver and driven units on the OD surface of the coupling. Place shims at all driver support points equal in thickness to the measured height difference.

3. **Horizontal Face and OD Alignment.** Using a straightedge, check alignment of ODs at sides of the coupling. Adjust driver as necessary to align the ODs and to set the gap equal at the sides. Don't disturb shims.

Temperature Compensation

To compensate for temperature difference between installation conditions and operating conditions it may be necessary to set one unit high, or low, when aligning. For example, centrifugal pumps handling cold water, and directly connected to electric motors, require a low motor setting to compensate for expansion of the motor housing as its temperature rises. If the same units were handling liquids hotter than the motor operating temperature, it might be necessary to set the motor high. Manufacturers' recommendations should be followed for initial setting when compensation for temperature change is made at cold installation.

Final alignment of equipment with appreciable operating temperature difference should be made after it has been run under actual operating conditions long enough to bring both units to operating temperatures.

7. ELECTRICAL POWER DISTRIBUTION

Electrical power distribution systems are composed of three main parts: the electrical service, distribution equipment, and feeder circuits. As shown in Fig. 7-1, the initial feed of power into the building or service comes through the electrical service (the power typically provided by a utility company). The service power is fed through service panels, which contain appropriate overcurrent protection devices. From there, power is distributed to additional overcurrent power distribution panels through feeder circuits. In most cases, individual conductors in raceways are used for these circuits, although busways are frequently used as well.

Services

It is important that electrical services be properly installed. This is true not only because the service carries all of the power for a facility, but also because the service conductors may be effectively *unfused*. For example, the typical service conductors to a small manufacturing facility are often fed directly from utility lines that are fused at several thousand amperes. In this case, the service conductors have so high a level of available current that they would melt and burn long before the overcurrent device would interrupt the current. Utility companies work at avoiding problems like this, but the problem is widespread nonetheless.

For these reasons, it is doubly important that you pay special attention to the installation of electrical services.

Service Requirements

Article 230 of the National Electrical Code covers virtually every aspect of electrical services.

Service Conductors

Service conductors are not allowed to pass through a building or structure and then supply another building or structure unless they are encased in two or more inches of concrete.

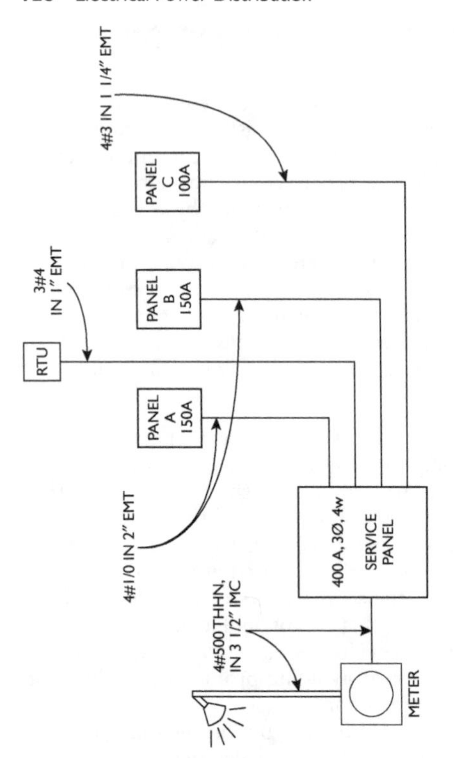

Fig. 7-1 Power distribution system.

Conductors are considered not to be in a building (although they actually are) in any of the following circumstances:

1. If they are encased in two or more inches of concrete.
2. If they are encased in a raceway, then enclosed in two inches of brick.
3. If they are in proper transformer vaults.

The only conductors other than service conductors allowed in service raceways are grounding conductors and load management conductors that have overload protection.

When a service raceway enters from underground, it must be sealed to prevent the entrance of gas. Empty raceways must also be sealed.

Service cables without an overall jacket must be at least three feet from windows or similar openings, except that they are allowed with less clearance over windows (rather than next to them).

Service conductors must have enough ampacity to carry the load placed upon them.

Service conductors are not allowed to be smaller than No. 8 copper or No. 6 aluminum. (When services feed only limited loads of single branch circuits, No. 10 copper [equivalent to No. 12 hard-draws, which the Code actually specifies] can be used.)

The size of the neutral conductor for a service must be at least the following:

1. *1100-kcmil or smaller service conductors:* The neutral must be at least as large as the grounding electrode conductor shown in *Table 250.66* of the NEC.
2. *Larger than 1100-kcmil service conductors:* The neutral must be at least 12½ percent of the size of the largest phase conductor. If parallel phase, the neutral must be 12½ percent of the equivalent cross-sectional conductor area.

Each service can have only one set of service conductors, except that multiple tenants or occupants of a single building can each have their own service-entrance conductors.

One set of service conductors is allowed to supply a group of service-entrance enclosures. Service-entrance conductors must be of sufficient size to carry their load.

Service conductors cannot be spliced, except as follows:

1. Clamped or bolted connections in meter fittings are allowed.
2. Service conductors can be tapped to supply two to six disconnecting means grouped at a common location.
3. A connection is allowed at a proper junction point where the service changes from underground to overhead.
4. A connection is allowed when service conductors are extended from an existing service drop to a new meter, then brought back to connect to the service-entrance conductors of an existing location.
5. Sections of busway are allowed to be connected to build the service.

Service Clearances

The importance of service clearances is safety. The chief dangers are events such as vehicles striking service drops (causing power outages and perhaps electrocutions) and people tampering with the service drops or having accidental contact with tools, machinery, or other objects. The intent of these requirements is to eliminate any such hazards (see Fig. 7-2).

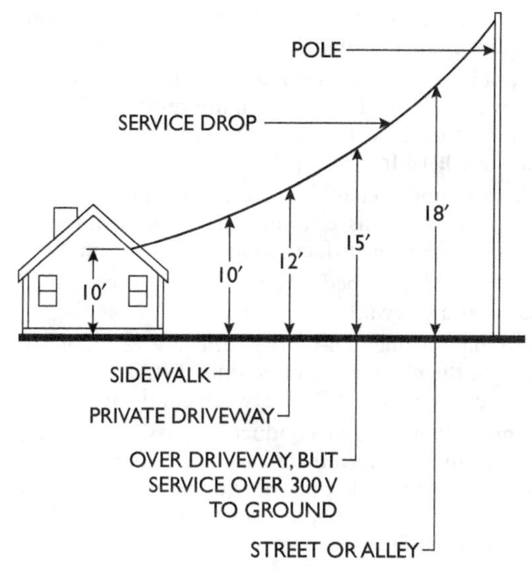

Fig. 7-2 Service clearance.

Service conductors not over 600 V must have the following clearances over grade:

1. Above finished grade, sidewalks, platforms, and the like from which the conductors could be reached by pedestrians (but not by vehicles) and where the voltage is not more than 150 V to ground: 10 ft.
2. Over residential driveways and commercial areas not subject to truck traffic and where the voltage is not more than 300 V to ground: 12 ft.

3. For areas in the above classification (12-ft rating) when the voltage is greater than 300 V to ground: 15 ft.
4. Over public streets, alleys, roads, parking areas subject to truck traffic, driveways on nonresidential property, and other land traversed by vehicles (orchards, grazing, etc.): 18 ft.
5. Conductors not over 600 V must have an 8-ft clearance over roofs. This clearance must be maintained with 3 ft of the roof surface, measured horizontally.
6. If a roof is subject to pedestrian traffic, it is considered the same as a sidewalk.
7. If a roof has a slope of 4-in. rise for every 12 in. of run or greater, the clearance can be only 3 ft. The voltage cannot be more than 300 V between conductors.
8. If no more than 4 ft of conductors pass over a roof overhang and are terminated by a through-the-roof raceway or approved support and the voltage is not more than 300 V between conductors, only 18-in. clearance is required (see Fig. 7-3).
9. Horizontal clearance from signs, chimneys, antennas, and the like need be only 3 ft.
10. When these conductors are attached to a building, they must be at least 3 ft from windows, fire escapes, and the like (see Fig. 7.4).

The point of attachment of service conductors to a building must be no lower than the above-mentioned clearances, but never less than 10 ft.

Service masts used to support service drops must be of sufficient strength or be supported with braces or guys (see Fig. 7.5).

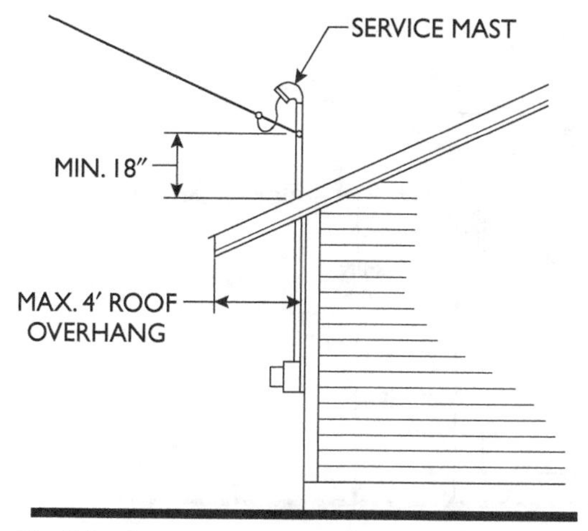

Fig. 7-3 Service mast installation.

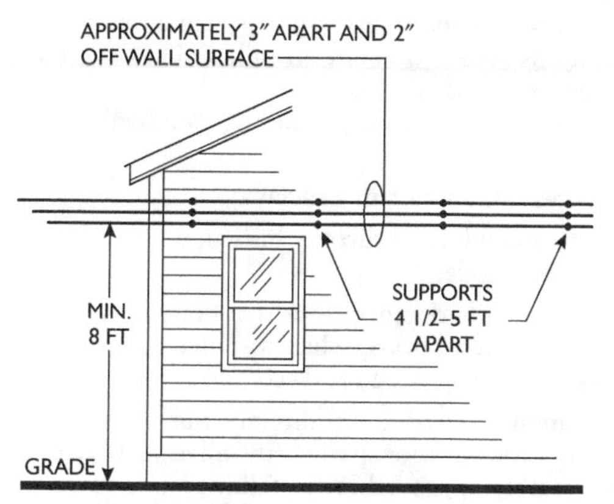

Fig. 7-4 Service run on wall surface.

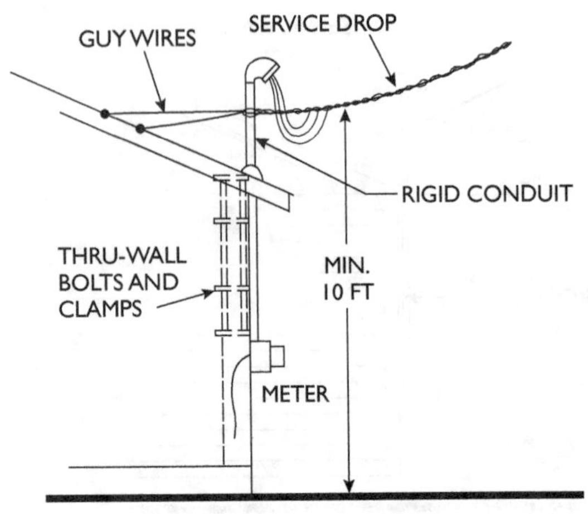

Fig. 7-5 Mast support.

Underground Service Conductors

Underground service conductors must be suitable for the existing conditions where they are installed. They must be protected where required (see Fig. 7-6).

The following bare service grounding conductors are allowed:

1. Bare copper conductors in a raceway.
2. Bare copper conductors directly buried, when soil conditions are suitable.
3. Bare copper conductors directly buried, without regard to soil conditions, when installed as part of a cable assembly approved for direct burial.
4. Bare aluminum conductors directly buried, without regard to soil conditions, when installed as part of a cable assembly approved for installation in a raceway or direct burial.

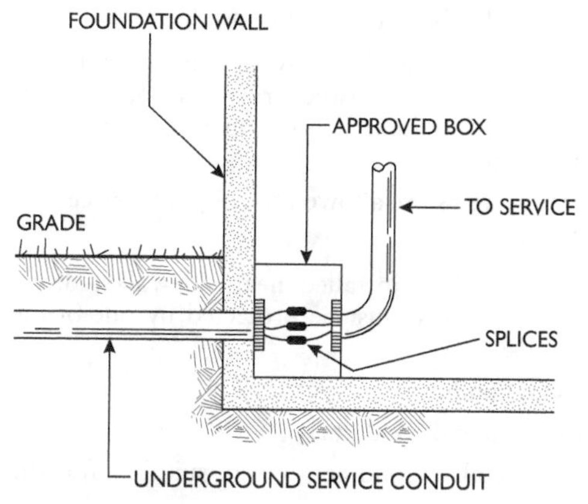

Fig. 7-6 Splicing underground service conductors.

Wiring Methods for Services

Service conductors 600 V or less can be installed using any of the following methods:

1. Rigid metal conduits
2. Intermediate metal conduits
3. Electrical metallic tubing
4. Service-entrance cables
5. Wireways
6. Busways
7. Cablebus
8. Open wiring on insulators
9. Auxiliary gutters
10. Rigid nonmetallic conduits
11. Type MC cables
12. Mineral-insulated, metal-sheathed cables

13. Liquid-tight, flexible, nonmetallic conduits
14. Flexible metal conduits, but only for runs of 6 ft or less between raceways or between raceways and service equipment. An equipment bonding jumper must be run with the conduit.
15. Cable tray systems are allowed to support service conductors.

Service-entrance cables installed near sidewalks, driveways, or similar locations must be protected by one of the following methods:

1. Rigid metal conduits
2. Intermediate metal conduits
3. Rigid nonmetallic conduits, when suitable for the location
4. Electrical metallic tubing
5. Other approved methods as specified in the code

Service-entrance cables must be supported within 12 in. of every service head, gooseneck, or connection to a raceway or enclosure. They must be supported at intervals of no greater than 30 in.

Individual open conductors must be mounted on insulators or insulating supports and supported as shown in *Table 230.51(C)*.

Cables that are not allowed to be installed in contact with buildings must be mounted on insulators or insulating supports and must be supported every 15 ft or less. They must be supported in such a way that the cables will have no less than 2 in. of clearance over the surfaces they go over.

Services must enter exterior walls with an upward slant so that water will tend to flow away from the interior of the building. Drip loops must be made (see Fig. 7-7).

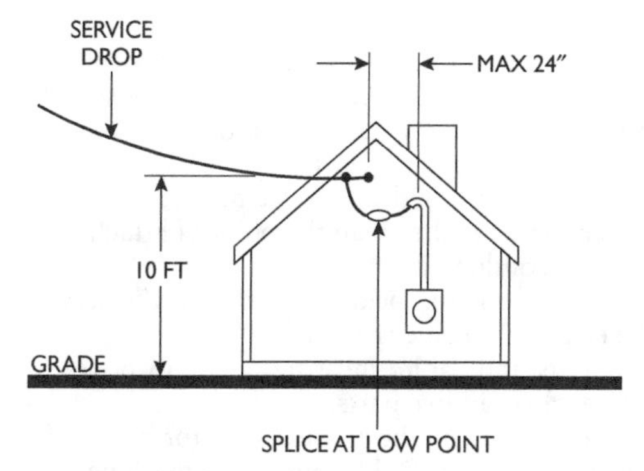

Fig. 7-7 Drip loop.

Service raceways exposed to the weather must be rain-tight and must be arranged so that they will drain.

Service raceways must have a rain-tight service head where they connect to service drops (see Fig. 7-8).

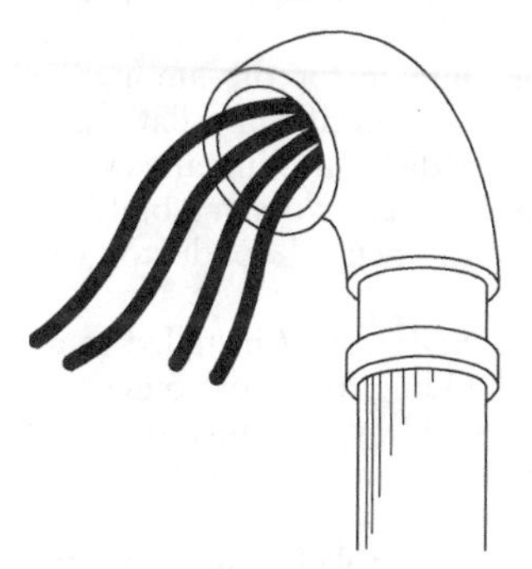

Fig. 7-8 Service head.

Service cables, unless they are continuous from a pole to the service equipment, must be provided with a service head or shaped into a gooseneck. When shaped into a gooseneck, the cable must be taped and painted or taped with self-sealing, weather-resistant thermoplastic.

Except where it is not practical, service heads for service-entrance cables must be higher than the point of attachment of the service drop conductors.

Drip loops must be made below the level of the service head or at the end of the cable sheath.

Service raceways and cables must terminate in boxes or enclosures that enclose all live parts.

For a 4-wire delta system, the phase conductor having the higher voltage must be identified by an orange marking.

Service Equipment

Energized parts of service equipment must be enclosed and guarded.

A reasonable amount of working space must be provided around all electrical equipment. Generally, the minimum is 3 ft. *Table 110.26(A)(1)* of the NEC contains more specific requirements.

Service equipment must be suitable for the amount of short-circuit current available for the specific installation.

A means must be provided for disconnecting all service-entrance conductors from all other conductors in a building. A terminal bar is sufficient for the neutral conductor (see Figs. 7-9 through 7-11).

The service-disconnecting means must be installed in an accessible location outside the building. If the disconnecting means is to be installed inside the building, it must be at the nearest possible point to where the service conductors enter the building.

All service disconnects must be suitable for the prevailing conditions.

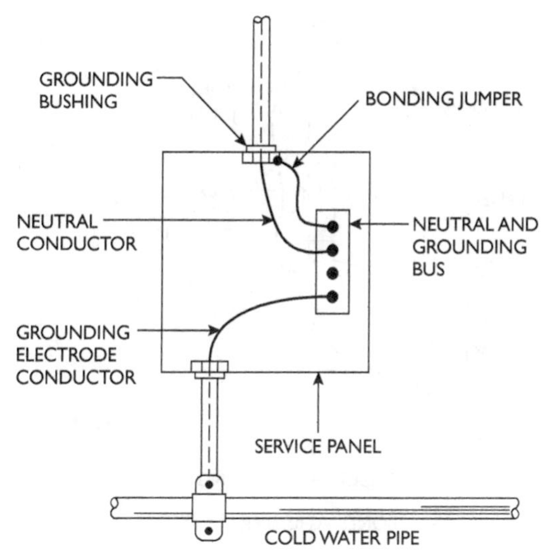

Fig. 7-9 Bonding the service panel.

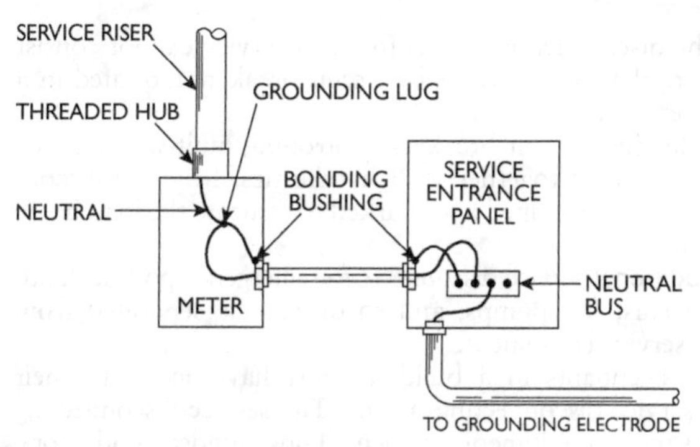

Fig. 7-10 Bonding the meter and service panel.

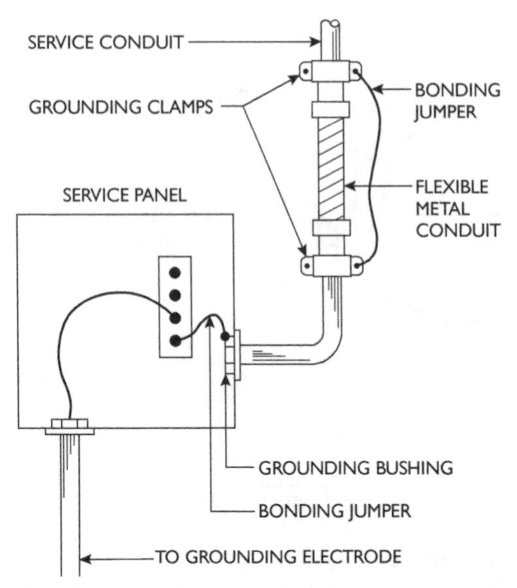

Fig. 7-11 Bonding around flexible service conduit.

The disconnecting means for each service cannot consist of more than six switches or circuit breakers mounted in a single enclosure (see Fig. 7-12).

Individual circuit breakers controlling multiwire circuits must be linked together with handle ties. Multiple disconnects must be grouped and marked to indicate the load being served.

Additional service disconnects for emergency power, standby systems, fire pumps, and so on can be separated from other service equipment.

All occupants in a building must have access to their own service-disconnecting means. The service-disconnecting means must simultaneously open all ungrounded conductors.

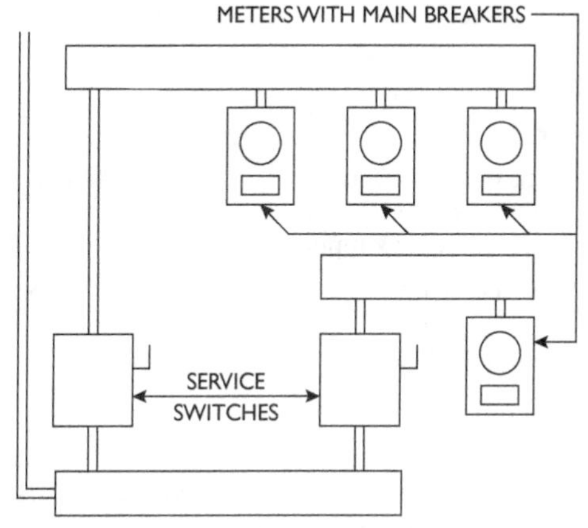

Fig. 7-12 Four service disconnects (up to six permissible).

The service-disconnecting means can be operable either manually or with a separately supplied power assist. This is a common feature of very large circuit breakers. When it is power operable, a manual override for use in the event of a power failure must be made possible.

The disconnecting means must have a rating no less than the load being carried.

Service-disconnecting means for one-family dwellings must have a minimum rating of 100 amperes; service disconnects for all other occupancies must have at least a 60-amp rating.

Smaller service sizes are permissible for limited loads not in occupancies. For loads of two 2-wire circuits, No. 8 copper or No. 6 aluminum conductors can be used. For loads with one 2-wire circuit, No. 12 copper or No. 10 aluminum conductors can be used. Service conductors may never be smaller than the branch-circuit conductors.

Only the following items are permitted to be connected to the line side of service disconnects:

1. Cable limiters or current-limiting devices.
2. Meters operating at no more than 600 V.
3. Disconnecting means mounted in a pedestal and connected series with the service conductors located away from the building being supplied.
4. Instrument transformers (current or potential transformers).
5. Surge protective devices.
6. High-impedance shunts.
7. Load management devices.
8. Taps that supply load management devices, circuits for emergency systems, fire pump equipment, stand-by power equipment, and fire and sprinkler alarms that are provided with the service equipment.
9. Solar photovoltaic or other interconnected power systems.
10. Control circuits for power operated disconnects. These must be provided with their own overcurrent protective and disconnecting means.
11. Ground-fault protective devices that are part of listed equipment. These must be provided with their own overcurrent protective and disconnecting means.

When more than one building or structure on the same property is under single management, each structure must be provided with its own service-disconnecting means (see Figs. 7-13 through 7-16).

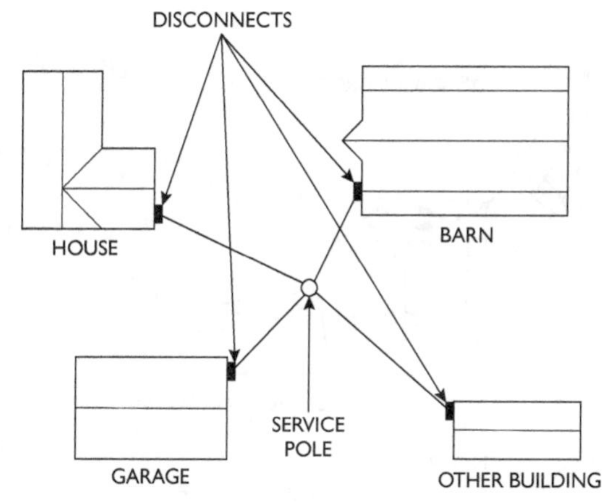

Fig. 7-13 Several buildings fed from one service pole.

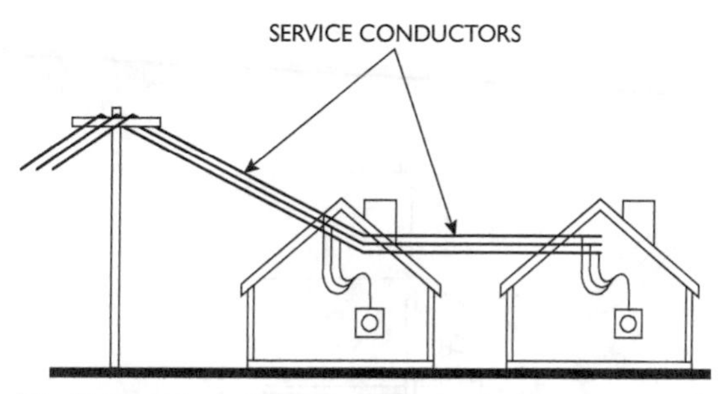

Fig. 7-14 One service drop to two buildings.

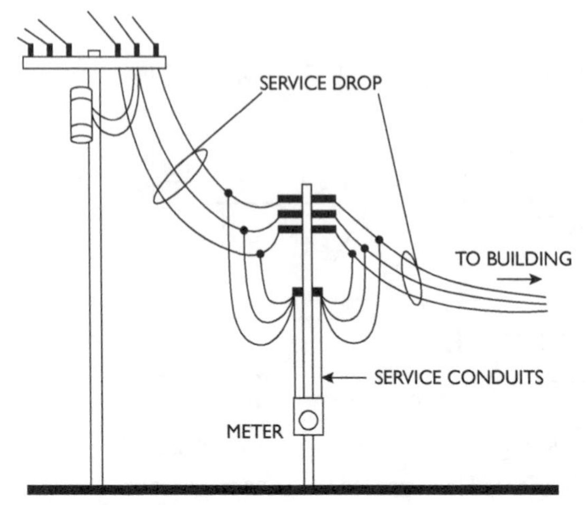

Fig. 7-15 Service with remote meter.

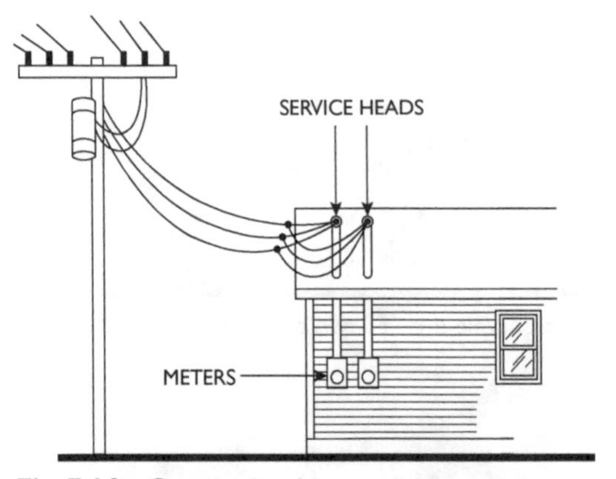

Fig. 7-16 One service drop serving two meters.

Overcurrent Protection for Services

An overcurrent protective device must be installed in each conductor and must be rated no higher than the ampacity of the service conductor.

No overcurrent device that would cause a safety hazard is allowed in a grounded conductor.

Services of 100 amps or more for solidly grounded wye systems must have ground-fault protection.

All occupants in a building must have access to their own service overcurrent protective devices.

Where necessary to prevent tampering, an automatic overcurrent protective device supplying only a single load can be locked.

The overcurrent protective means must protect all equipment except the following:

1. Cable limiters or current-limiting devices
2. Meters operating at no more than 600 V
3. Instrument transformers (current or potential transformers)
4. Surge protective devices
5. High-impedance shunts
6. Load management devices
7. Taps that supply load management devices, circuits for emergency systems, fire pump equipment, stand-by power equipment, and fire and sprinkler alarms that are provided with the service equipment
8. Solar photovoltaic or other interconnected power systems
9. Control circuits for power operable disconnects. These must be provided with their own overcurrent protective and disconnecting means.
10. Ground-fault protective devices that are part of listed equipment. These must be provided with their own overcurrent protective and disconnecting means.
11. The service-disconnecting means

Services at Greater Than 600 V

Services that operate at over 600 V are far more hazardous than services that operate at lower frequencies. Therefore, the requirements for these services are quite a bit more stringent. Safer and heavier methods of wiring are required, as is more mechanical protection. The requirements for these services are given below. Note that these requirements are in addition to, and in some cases modify, the other requirements of *Article 230*.

Service conductors over 600 V can be installed using any of the following methods:

1. Rigid metal conduits
2. Intermediate metal conduits
3. Service-entrance cables
4. Busways
5. Cablebus
6. Open wiring on insulators
7. Rigid nonmetallic conduits
8. Cable tray systems that can support service conductors

Conductors and supports must be of sufficient strength to withstand short circuits.

Open wires must be guarded.

Cable conductors emerging from a metal sheath or raceway must be protected by potheads.

Feeders

Feeders are circuits that distribute fairly large quantities of power between distribution panels (or service panels) and branch-circuit panels. They are among the most expensive parts of a wiring system because they require large and expensive conduits and conductors.

The physical routing of feeders is an important consideration on the job site. Since feeders can easily cost $100 to $200 per foot, the length of a run is a rather serious matter and should be minimized wherever possible.

The requirements for feeders are found in *Article 215* of the National Electrical Code and are generally as given below.

The Sizes and Ratings of Feeders

Most feeders are rated 100 amps or higher. It is permissible, however, for feeders to be rated less than this for certain uses.

Feeder conductors must have an ampacity no less than that required to carry their load.

Feeders can be no smaller than 30 amperes when the load being supplied consists of the following types of circuits:

1. Two or more branch circuits supplied by a 2-wire feeder.
2. More than two 2-wire branch circuits supplied by a 3-wire feeder.
3. Two or more 3-wire branch circuits supplied by a 3-wire feeder.
4. Two or more 4-wire branch circuits supplied by a 3-phase, 4-wire feeder.

Feeders must be protected with overcurrent protective devices.

Grounding and Other Requirements

Since feeders carry large amounts of current, they can supply large fault currents. Therefore, the grounding requirements are especially important. Many feeders could supply fault currents of several hundred amps without tripping their circuit breaker. If there were no good grounding system, these very large fault currents could exist (causing great damage and danger) without ever being detected.

Feeders that contain a common neutral are allowed to supply two or three sets of 3-wire feeders or two sets of 4- or 5-wire feeders.

In any metal raceway or enclosure, all feeders with a common neutral must have their conductors run together.

When a feeder supplies branch circuits that require equipment-grounding conductors, the feeder must supply a grounding means to which the equipment-grounding conductors (from the branch circuits) can be connected (see Figs. 7-17 and 7-18).

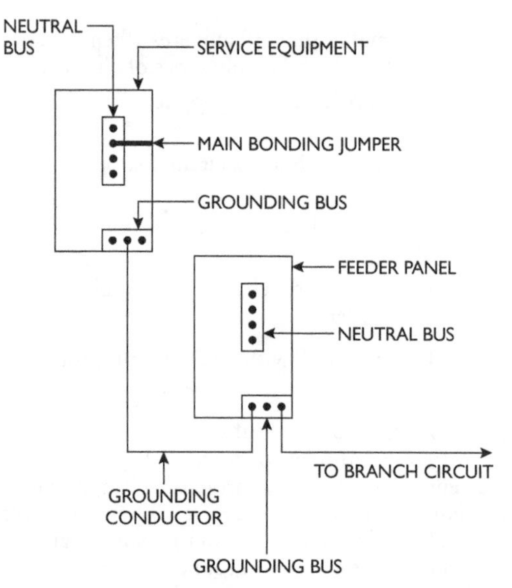

Fig. 7-17 Grounding of feeder panels.

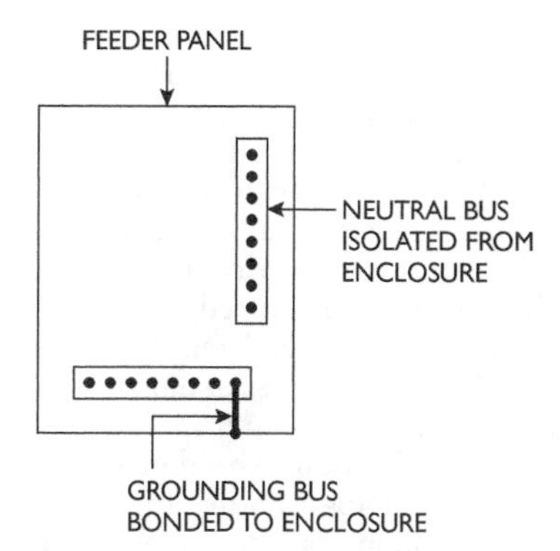

Fig. 7-18 Bonding a grounding bus.

Two-wire ac or dc circuits of two or more ungrounded conductors are allowed to be taped from ungrounded conductors of circuits that have a grounded neutral. Switching devices in each tape circuit must have a pole in each ungrounded conductor.

For a 4-wire delta system, the phase conductor having the higher voltage must be identified by an orange marking.

Feeders that supply 15- or 20-ampere branch circuits that require ground-fault protection can be protected against ground faults, rather than protecting individual circuits or receptacles.

Panelboards

Historically, the delivery of large panelboards, switches, and similar service equipment has been rather slow. Therefore, the electrical personnel should see to it that the equipment is ordered shortly after the work has been ordered and then follow up on the order to make sure the equipment is delivered on time.

It is especially important that the panelboard cabinets ("cans," as they are often called) are delivered to the job site in plenty of time. In most cases these cabinets must be installed during the time the walls are erected and all conduits must be run and attached to these cabinets prior to closing in the building. If there is an anticipated delay on the complete panelboards, the electrical supervisor should obtain a guarantee that the cans will be shipped at a specified date. Then the interiors may be installed later.

When ordering service equipment, its physical dimensions with relation to doorways, beams, ceilings, and space available must be given careful consideration. The dimensions may be checked by reference to catalog information in the case of standard equipment or to shop drawings prepared by the manufacturer in the case of custom-built equipment.

Weight must also be given consideration in relation to handling, moving equipment over floors, floor covering, and installation at a point above the floor level. While the physical dimensions of such equipment may be such that it could be moved into place as a completely assembled unit, the weight of the complete assembly is often such that it has to be handled in sections.

In some instances, it may be necessary to move such equipment over floors that are not of sufficient strength to bear the weight load of the complete assembly or over floor coverings that would be damaged in moving equipment over it. In such instances, the equipment has to be handled in sections or else adequate provision has to be made for additional support or protection of the floor surface.

Sometimes, particularly in industrial plants, you may be required to install such equipment at some height above the floor level. The weight of the equipment as well as its physical dimensions must be given consideration in arranging for installation facilities and in determining the adequacy of the permanent supporting structure. It may be necessary to provide supporting structures of greater strength than was originally designed in order to prevent the possibility of the failure of the supporting structure and resultant damage to the equipment and injury or death to humans.

You should also be concerned with the preceding condition if any changes have been made in the original working drawings. For example, during a plant addition in which heavy electrical equipment is required to be installed at some distance from the floor and to be supported on the building's structural members, the plant maintenance supervisor may request that the equipment be relocated to a different area for easier access. This slight move may not appear to cause any problems, but always check with the architect or engineer in charge of the project to determine if the supports are adequate. If the plant's own engineering department has designed the system and there is any doubt in the electrical supervisor's mind about the possibility of failure of the supporting structure, it is wise to commission a registered structural engineer to examine the situation.

When panelboards are received, packing boxes should be immediately examined for damage, and if damage exists, it should be pointed out to the truck driver, who can report the fact to the trucking company. Boxes should be stored in a dry and clean location until panels are installed. When it is not possible to install fronts of panels at the same time as the interiors, the fronts should be left in the packing boxes until they are installed. The installation of panelboards, service equipment, and load centers is a specialty job. Experienced mechanics should be assigned to this type of work.

When boxes are located on the surface of existing walls, either the walls should be flat or the low points should be shimmed out so that boxes are not strained out of shape when secured to walls. For flush boxes, the boxes must be securely fastened and be made both plumb and flush with finished wall surfaces.

No panelboard interiors should be installed until all of the wires have been pulled in and the plastering finished around the box. Before panelboard interiors are installed, the plaster and building grit should be cleaned off the wires and the front fasteners and out of the box.

Panelboards should be carefully centered and properly plumbed. Panelboards must also be adjusted outwardly so that the opening in the front makes a close fit with the edges of the panelboard. At the same time, the front must be flat against the finished plaster or wall. As a final step, directory cards on the inside of panel covers should be marked with the locations of all branch circuits.

While the use of metal-enclosed panelboards is increasing, assemblies of individual enclosed externally operated switches or circuit breakers, interconnected by nipples or gutter connectors, are often used.

Such assemblies are normally mounted on a backboard of wood sheathing. Some will require the backboard to be painted or covered with sheet metal prior to mounting the equipment units on it. In special cases, the specifications may require that the assembly be mounted on steel plate or on an angle iron or channel iron structure.

In most cases, the general arrangement of the equipment and its space requirements should be given consideration before the installation begins. Most contractors prefer to make a scale drawing of the layout to get the best arrangement and make certain the components will fit into the space provided. Several schemes should be tried in order to determine the best arrangement with respect to neatness and efficiency.

Once the general arrangement has been laid out, the backboard should be constructed out of channel iron, angle iron, plywood, or a combination of these materials. The equipment is then secured to the backboard and interconnecting nipples, connectors, wire trough, and so on, are installed as required.

The feeder conductors may then be pulled into the trough and nipples and to the disconnect switches so that terminations may be completed. At this time, each feeder should be identified and labeled. Install proper size fuses; check for ground faults; and the system is ready to be energized, at least to the load side of the switches.

Power Factors

An ac electrical system carries two types of power. The first is correctly called *real* (or *true*) *power,* and the other is called *apparent power.* True power is measured in watts and is the power that drives the system's load. Apparent power is measured in volt-amps and is the power required to support magnetic fields in inductive electrical equipment, typically electric motors and fluorescent ballasts.

Real power is always less than apparent power in reactive circuits. If you refer to Fig. 7-19, you will see two triangle diagrams. In the bottom drawing, you see the relationship between real power, apparent power, and VAR (volt-amperes reactive). The ratio of real power to apparent power is called the *power factor*. A perfect power factor (where there is no difference between real and apparent power) is 1.0, which is called *unity*. Typical power factors for industrial buildings would be .9 or less.

Apparent power is the power that you can measure with a typical voltmeter and ammeter. The technical definition is the measure of alternating current power that is computed by multiplying rms (root-mean-square) voltage and rms current. This is called apparent power, because it is the power that is measured by the most common means—the power that appears to you.

Real power is more complex. In a purely resistive circuit (with no reactance), apparent power is the same as true power. But when voltage and current move out of phase with each other, things get complicated. To measure real power, the waveform must be divided into many, many segments, and a very large number of voltage and current readings must be taken, and then averaged. Specialized watt meters and reactive power meters are designed to do this.

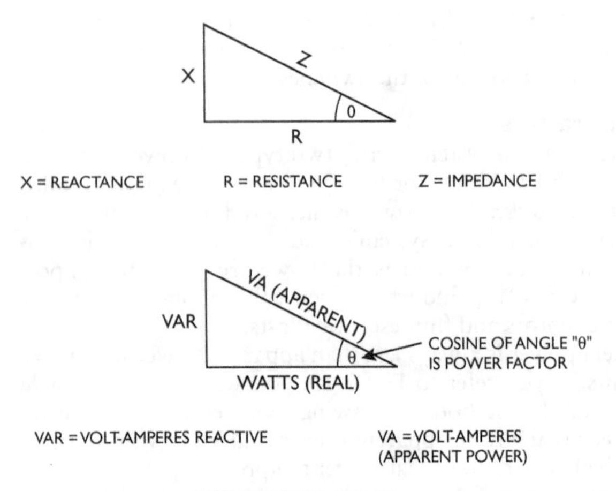

Fig. 7-19 Power factor diagrams.

Power factor is a major concern to both utility companies and large factories. Because facilities such as factories use a great deal of inductive equipment, their power factors are substantially below unity. Thus, the utility company must oversize its power production and distribution systems. Because electric meters don't read apparent power, the factory cannot be directly charged for the apparent power it draws from the utility lines. So, in order to be paid for all the power they provide, the utilities check the power factor of the factories and bill the factories a separate power factor charge, which can be substantial.

In order to avoid the high power factor charges, factories often install *power factor correction equipment.* Since power factors are created by inductive reactance, the method of correcting them is to introduce capacitive reactance into the circuits. This has the effect of canceling the inductive reactance.

(See Chapter 1 for a related discussion of leading and lagging voltages.) The two most popular methods of putting capacitive reactance into the circuits are capacitor installations and the use of synchronous motors. (See Chapter 4 for a detailed discussion of synchronous motors and their use in power factor correction.) Capacitors can be installed either directly at the individual inductive loads themselves (such as motors) or in a large grouping.

Dirty Power

Because of the sensitive nature of the electronic circuits in data-processing equipment, current and voltage problems that are never noticed in the operation of other electrical machines can cause great problems to data-processing equipment. If you have ever taken a good look at an integrated circuit chip, you understand why a very small voltage surge could damage it. The distances between components in the chip are so small that a very slight *spike* could quite easily jump across. Thus, these processors must be protected from erratic electric power.

Power problems can generally be classified into the following three types:

1. **Transient voltages and oscillatory overvoltages.** Typical causes for these problems are utility company switching, lightning, or operation and switching of certain loads within the building. Most typically, problems arise from the energizing or deenergizing of inductive loads such as motors or fluorescent lighting. These items operate with a magnetic field set up around their various inductive coils (ballast transformers or motor windings). When power is removed from these loads, their magnetic fields collapse, inducting a momentarily high voltage (frequently at least several times line voltage) back into the coil. This voltage is thereby fed into the building's general wiring system and could affect any piece of electrical equipment connected to the wiring system.

2. **Momentary overvoltage or undervoltage.** These are longer-term changes (typically between 4 and 60 cycles) in voltage to the data-processing equipment. Often these conditions are caused by changes in the voltage supplied by the utility company. Undervoltages are most commonly caused by large additions either to the utility system's load (such as when distribution circuits are switched onto the system) or to the building's load (such as when a large motor or other large load is switched on). These are the changes you notice when the lights momentarily dim. Overvoltages are usually caused when the utility lines are less loaded than normal, especially when a large load drops out suddenly. Frequently, overvoltages are encountered at night when the utility company's load is lighter than usual.
3. **Power outages.** Outages occur when large faults, such as lightning strikes that blow distribution fuses, occur. Utility or on-site malfunctions may also cause such outages.

Variations

In general, computer equipment can accept voltages within the tolerances that are set for utility company power, although a few types of computer products call for more stringent requirements. However, this voltage must remain constant; not many computers will tolerate a voltage loss of more than about 30 milliseconds. Most computers have a frequency tolerance of about 0.5 Hz, which is rarely problematic.

Power variations beyond acceptable limits can cause errors in calculations, output errors, loss of data, unscheduled shutdowns (a fancy name for "crashes"), downtime, and need for repairs to equipment (an important concern to data-processing managers).

One of the most extensive studies of utility power ("The Quality of U.S. Commercial AC Power" by M. Goldstein and P. D. Speranze, IEEE April, 1982) outlined the most common power line disturbances and their frequencies. All power line problems were broken down into the various percentages of occurrence. The findings follow:

Voltage sags, 87.0%

Impulses (transient spikes), 7.5%

Power failures, 4.7%

Surges, 0.8%

Solutions to Power Problems

There are several basic technologies (with many modifications) that are used to modify *raw* utility power before delivering it to the data-processing equipment. The technologies can, in general, be divided into two categories: power conditioners and power synthesizers.

Power conditioners modify and improve the incoming power waveform by clipping, filtering, increasing, or decreasing the voltage or by like modifications. These conditioners are placed between the utility power and the computer equipment.

On the other hand, power synthesizers use the incoming utility power strictly as an energy source to power the creation of a new, separately derived power supply. The synthesized power is designed to be fully within the requirements of the computer equipment.

Power modification is commonly provided by transient suppressors, voltage regulators, isolating transformers, or some combination of these devices.

Power synthesis may be provided by static electronic semiconductor inverters (UPS systems), special motor-generator sets, or magnetic synthesizers.

No one type of power conditioner or synthesizer is best for all situations. Selecting the proper computer power equipment requires a carefully considered decision based upon a good knowledge of the various types available.

Computer power equipment tends to be applied according to the risk involved in a given application. For instance, a power outage normally causes no great loss to a home computer; therefore, the product normally used is some inexpensive type of surge suppressor. But for the computer system for a large bank, no amount of protection short of the best available will do; the bank simply has too much to lose if the system should lose some of its data. The question is, Where do the risk curve and the cost curve cross? The answer will be different for every system. And often, no two people assess the risk identically. There is no concrete answer for how much risk is involved. It is strictly a *judgment call.*

Surge Suppressors

Surge suppressors are designed to do the following:

1. During normal operations, they can't interfere with the circuits they protect.
2. The clamping voltage (the voltage at which the suppressor will connect the line with excess voltage to ground) must not be greater than the surge-withstand rating of the equipment it protects.
3. Clamping speed must be fast enough to prevent damage to the protected equipment.
4. The suppressor must be capable of withstanding surges without damage.

Types of surge suppressors are as follows:

1. **Spark gaps.** A spark-gap surge suppressor consists of two electrodes placed in free air or some other arc-quenching material. When too high a voltage is encountered, it jumps the air gap (in the form of a spark) and is directed to ground, thus removing it from the power lines. Spark-gap arrestors are often used for removing large lightning surges from the

power lines. One problem with spark-gap devices is that they often allow a follow-through current once an arc is established. This is usually compensated for by using a magnetic blowout, a circuit breaker, or a deionizer.

The advantages of spark-gap devices are simplicity and reliability, a long life expectancy, high energy-handling capacity, zero power consumption, very low voltage drop across gap during conduction, low capacitance, reasonably quick response time, and bipolar operation (works in either direction). The disadvantages of spark-gap devices are that their firing voltage is dependent upon atmospheric conditions and surge-rise time, that usage is limited to circuits of relatively high voltages, and that devices used alone won't extinguish follow-current.

A typical spark-gap arrestor is shown in Fig. 7-20. This is an inexpensive item and is primarily for removing lightning surges.

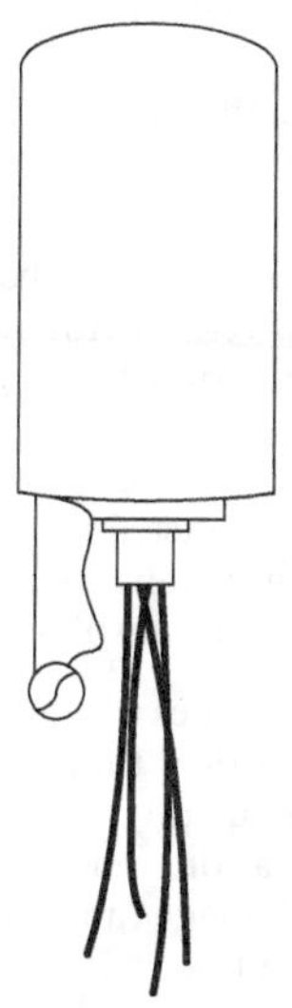

Fig. 7-20 Spark-gap arrestor.

2. **Gas tubes.** Gas tubes are similar in operation to the spark-gap arrestors. However, they eliminate the problems related to atmospheric conditions by enclosing the gap in an atmosphere of neon, argon, krypton, or a similar gas.

The advantages of gas tubes include a low cost, small physical size, good life expectancy, high current capacity, low clamping voltage, and a fairly low capacitance. The disadvantages of gas tubes are that follow-current must be limited on power circuits, that the firing voltage depends on surge rise time, that they don't absorb appreciable surge energy, and that they can be ionized by strong RF (radio frequency) fields.

3. **Metal oxide varistors.** These devices, commonly called MOVs, are made of sintered zinc oxide particles pressed into a wafer and equipped with connecting leads or terminals. These devices have a more gradual clamping action than either spark-gap arrestors or gas tubes. As the surge voltages increase, these devices conduct more heavily and provide clamping action. And, unlike spark-gap arrestors and gas tubes, these devices absorb energy during surge conditions.

Advantages are that they can be used for low-voltage applications, have a fast response time, absorb energy, and require no follow-current protection. The disadvantages are that their clamping time is dependent on the surge wavefront, that they have a limited surge life expectancy, that they fail partially shorted, that they require external fusing for power applications, and that they have a high capacitance.

4. **Silicon avalanche devices.** These devices are essentially zener diodes that are designed to handle large surge currents without suffering damage. The junction construction for these devices is typically 10 times larger than that of normal zener diodes. The junction is sandwiched between silver electrodes to improve current distribution and to aid in thermal dissipation. These devices have an extremely fast clamping action and an absolute clamping level.

Advantages of silicon avalanche devices are a high clamping speed (less than 1 nanosecond), a firm clamping threshold, small physical size, and availability in a bipolar configuration. The disadvantages are that high surges may cause damage and that lead length substantially affects clamping time.

Combinations

No one type of surge suppressor is clearly the best in all situations. It is possible, though, to combine these various devices to fit almost any application.

To explain the way multiple suppressors can be used together, imagine the path that a surge might follow on the way to the data processing equipment you want to protect.

Primary suppressor

The first lines of defense against a power-line lightning or switching surge are the primary suppressors installed by the power company. These devices are installed between each phase conductor and the grounded neutral. They are commonly located at step-down transformers and at poles serving underground distribution. These devices are designed to clamp surges to a value below the insulation breakdown rating of the utility company's transformers and cables.

Secondary suppressor

The second line of defense against surges is a suppressor located at the building's service entrance location. This device consists of a low voltage spark gap with a series thyrite element connected between each phase conductor and the grounded neutral. (The thyrite elements control the follow-current mentioned previously.) Other, more effective devices consist of a large MOV (metal oxide varistor) connected between the phase conductors and the grounded neutral. It is worth noting that these suppressors are not required by code for underground services even though the service may be fed from a nearby overhead pole.

Equipment suppressors

The term *equipment surge suppressors* refers to types of surge suppressors that are designed to protect pieces of equipment rather than the electrical wiring of an entire building. Equipment surge suppressors range widely in both complexity and cost. These factors are usually directly proportional to the clamping effectiveness of the device. Most of these systems require a suppressor at the building's electrical service entrance (usually a spark-gap arrestor) to prevent the equipment surge suppressor from being exposed to large transient voltages. The spark-gap arrestor can handle large surges while most equipment surge suppressors cannot. Nonetheless, these equipment surge suppressors may often encounter surges as high as 2000 V.

One of the simplest, most common, and least expensive forms of equipment surge suppressors consists of a varistor and a thermal fuse connected across the line and neutral conductors. These are commonly found in the plug or socket or in outlet strip devices containing surge suppression. For a 2000 V, 15-ampere surge, the equipment being guarded will typically be exposed to no more than 500 V.

For more sensitive equipment, hybrid devices are available to provide a lower clamping voltage and a very fast response time. These assemblies use high-energy metal oxide varistors for the first stage of protection, with the varistor and fuse connected between line and neutral. The line voltage then passes through a large air core inductor (coil) to a silicon *avalanche* second stage made of series-connected bipolar suppression diodes between line and neutral. The assembly works as follows: The MOVs absorb most of the surge while the diodes provide a very fast clamping time. The induction coil provides a large enough voltage drop during the surge to prevent damage to the diodes.

It is worth mentioning that all of these MOVs, diodes, and the like are relatively small electronic parts. While the exact construction of each varies from one manufacturer to another, they are almost always small disks or cylinders with two leads.

Voltage Regulators

Voltage regulators are used to alleviate problems with overvoltage or undervoltage. The voltage regulator, as shown in Fig. 7-21, is basically a 1:1 transformer with a variable tap on the secondary side. A special voltage-sensing circuit monitors the voltage on the secondary side and adjusts the variable tap to ensure that the voltage stays within the proper limits.

A voltage regulator should always be used with a surge suppressor (some models have them built in). When so designed, this device goes one step further than surge suppression. With a voltage regulator properly installed, all over- or undervoltages (except extremely severe ones) are adjusted before they reach the data-processing equipment.

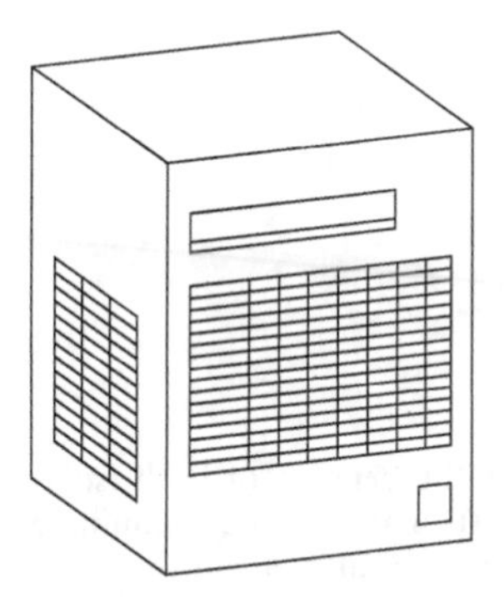

Fig. 7-21 Voltage regulator.

Voltage regulators are quite a bit more expensive than surge suppressors, but they remain within the budget of most companies and even most individual computer users. They are far less expensive than power synthesizers.

Isolation Transformers

Isolation transformers (as shown in Fig. 7-22) electrically isolate computer equipment from raw utility company power. The main advantage of these isolation transformers, some of which are shielded, is that they filter out very short electrical spikes, even those of several hundred volts. They are far less effective at filtering out longer-lasting spikes. The longer spikes on the primary windings will be transferred to the secondary windings as a higher than appropriate voltage, which is passed on to the equipment.

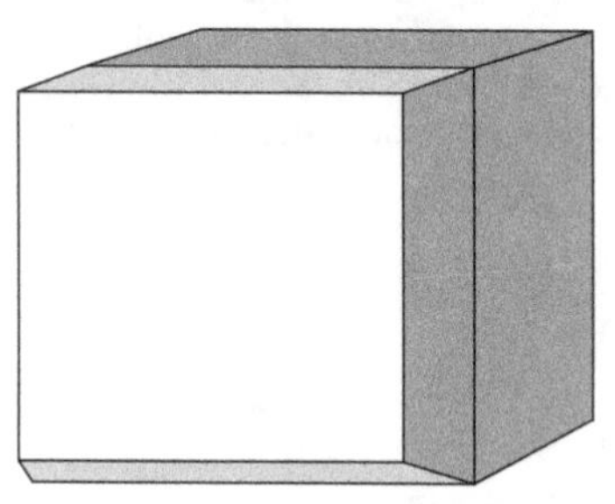

Fig. 7-22 Isolation transformer.

Isolator-Regulators

These combination devices offer the benefits of both isolation transformers and voltage regulators. They eliminate long-duration voltage fluctuations and also shield the data-processing equipment from spikes, voltage dips, and electrical noise.

The free-standing, prepackaged *computer power centers* are typically of this type of design (see Fig. 7-23). Some use ferroresonant transformers, and some use tap-switching transformers as the isolating means. Ferroresonant transformers are faster at eliminating spikes and surges but may also have less desirable effects when operating under a strongly inductive load (common in certain computer applications) and have been known to have phase-angle problems as well.

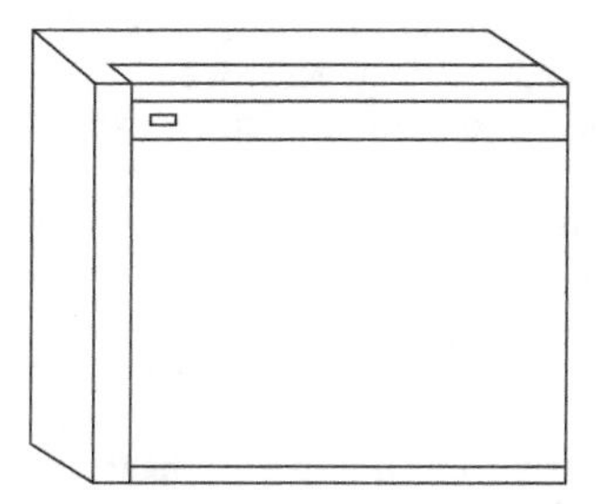

Fig. 7-23 Computer power center.

Ferroresonant transformers use a steel core that is partially magnetically saturated, which means that part of the transformer can handle no more magnetic flux. Thus, an increase in the voltage to above normal on the primary windings cannot cause an increase in the magnetic flux within the part of the transformer that is already saturated. As a result, the secondary voltage does not increase along with the spike on the primary side, as there is not enough magnetic flux to transfer it to the secondary windings.

Rotary UPS

A rotary UPS (uninterruptable power source) system is a motor-generator set that uses line power to turn a generator, which in turn provides a new source of ac power that is free from interference. This type of system is not a power conditioner as are the others covered up to now. The motor-generator is a power synthesizer. It provides a new and independent source of power, what the code calls "separately derived."

This type of system is significantly less expensive than the true UPS systems but will do everything that the true UPS systems do, with the exception of supplying power during power outages.

Other than being of a slightly higher quality, rotary UPS systems differ little from standard motor-generator sets. The motor may be either the synchronous or induction (sometimes called the asynchronous) type. These motor-generator sets are often incorporated into a computer power center, which means the set is inside a sound-deadening enclosure as well.

Rotary UPS systems come either as a single-shaft unit or as separate units. Single-shaft units have a common shaft between the two parts, decreasing the number of bearings required. Although this is good for maintenance, there may also be transient voltages traveling through the shaft from one side to the other. The separate units are connected with belts or similar means. Thus, no transients can be transferred, but maintenance is more difficult.

Several motor options exist. The synchronous motor is inherently tied to power-line frequency and runs the generator at the proper speed. It is more complex to start and takes time to restart in the event of a power outage. The induction motor is less expensive and easier to operate, but it doesn't run up to speed because of the "slip" inherent in this type of motor. One solution to this problem is to use the induction motor to drive the generator via pulleys and belts that are sized to run the generator slightly faster than the motor, thus maintaining proper speed.

Another motor option is to use a dc motor with two sources of power. The first source is utility company power, which is run through a rectifier band to produce dc power. The second is a battery bank connected in parallel with the rectified dc power. In the event of an outage, the battery bank will keep all the data-processing equipment running. The length of time that it will be kept operating depends upon the number of batteries used, but it is not too difficult to power the system long enough for an orderly shutdown.

A typical rotary UPS system is shown in Fig. 7-24.

UPS Systems

A UPS system is an assembly of electrical and electronic equipment that is designed to maintain the complete continuity of electrical power to critical loads, primarily to important data-processing machines.

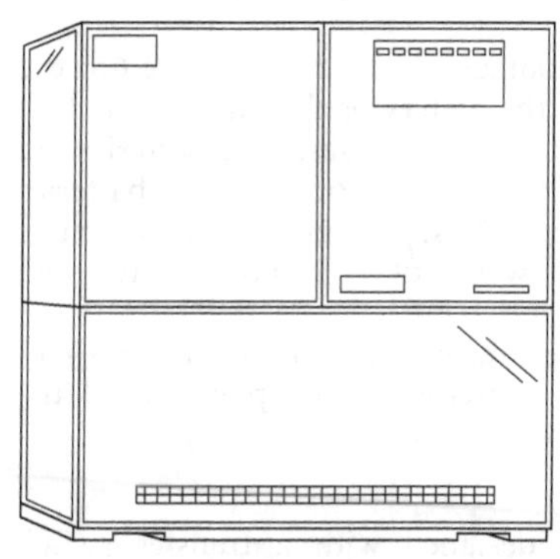

Fig. 7-24 Rotary UPS system.

The UPS system operates by using three basic devices:

1. ac to dc rectifier
2. Battery bank
3. dc to ac inverter

There are three modes of operation for these systems: normal mode, ac power failure, and bypass.

Normal mode

During normal operation, power flows into the rectifiers and is transformed into dc power. A small amount of this dc current is sent first to a battery charger and serves to maintain the batteries at full charge. The rest of the dc power goes to the dc to ac inverter, where it is synthesized into 60-cycle ac voltage. This power goes through a transfer switch and to the load being served.

When operating this way, a new voltage is being synthesized, using the utility company power only as an energy source. The power fed to the data-processing equipment is as pure as you can get, fresh from a new source.

ac power failure mode

In the event of a power outage, power is maintained to the processing equipment by the battery bank and inverter. This is done without any switching at all, since the batteries are always in the circuit. When power drops out, the batteries are already there and begin to supply power. The batteries will continue to supply power until utility power is restored or an orderly shutdown can be made.

Sometimes separate generators are also used. In the event of a power failure, the batteries supply power until the backup generator comes on line.

Bypass mode

Most UPS systems are designed with a transfer switch through which the inverter output flows to the load being served. These switches are designed to protect the load from interruption in the event of some type of failure in the UPS system.

The bypass source can be the utility line, a backup generator, another inverter, or the like. The transfer switch can be operated automatically or manually.

A properly designed UPS system (shown in Fig. 7-25) maintains power to a critical load at all times. The protection offered by such a system is unmatched, except perhaps by a rotary UPS system with a dc motor and a battery backup. There are, however, a few problems with these systems: (1) their high cost, (2) the large amount of space required, and (3) the maintenance of the batteries.

Computer Grounding

The problem with computer grounding is twofold:

1. Allowing erratic voltages to get into the data-processing equipment via the grounding system.
2. Providing a good high-frequency grounding path.

Stray ground voltages have caused the most problems for electricians. A computer's isolated ground does not have to terminate at electrical panels but must be connected at the same ground connection point as the normal system ground. (See *Section 250.146(D)*, of the NEC.) The purpose of the isolated ground is to prevent stray voltages from coming into computers via the building's grounding system. These voltages are usually caused by magnetic induction from other conductors sharing the same conduits with the grounding conductors (the reason for using dedicated lines). They are also caused by voltages being induced into the conduit system, which is commonly connected to the grounding wire. This connection gives these voltages access to the computer (the reason for isolated grounds).

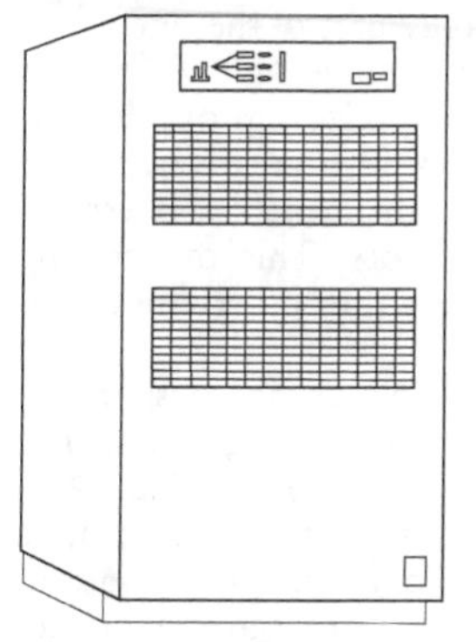

Fig. 7-25 UPS system.

The best way to cure these problems is to run dedicated lines (individual circuits run in individual conduits and boxes) and to provide isolated-grounding receptacles and an insulated, independent grounding system that ties directly to the termination of the grounding electrode conductor.

Harmonics

Let us begin here by defining this problem and the terms that will be used.

First of all, harmonics are electrical signals at multiple frequencies of the power line frequency. The idea is the same as that of harmonics in music. Harmonics are caused by many electronic devices. They don't cause problems to the computer equipment, but they do cause problems to the power distribution system. They are especially prevalent wherever there are large numbers of personal computers, adjustable-speed drives, and other types of equipment that don't use all of the sine wave but rather draw current in short pulses.

The Source of the Problem—Power Supplies

Many types of electronic equipment, particularly computers and the like, draw current during only part of the sine wave. These pieces of equipment are said to have *switched power supplies*. Although switched power supplies improve the efficiency of electronic equipment, they also create harmonic currents. Harmonic currents cause transformers to overheat, in turn overheating neutral conductors. This overheating may lead circuit breakers to trip when they otherwise would not.

A normal 60-cycle power line voltage appears on the oscilloscope as a sine wave, as shown in Fig. 7-26. The switched power supplies draw current as shown in Fig. 7-27. With the harmonics present, the waveform is distorted, as shown in Fig. 7-28. These waves are described as nonsinusoidal. The voltage and current waveforms are no longer simply related—and are called nonlinear. Nonlinear loads draw current in abrupt pulses that distort current wave shapes, causing harmonic currents to flow back into other parts of the power system.

In these power supplies (also called diode-capacitor input power supplies), the incoming ac voltage is rectified with diodes and is used to charge a large capacitor. After a few cycles, the capacitor is charged to the peak voltage of the sine wave (170 V for a 120-V ac line). The electronic equipment then draws current from this high dc voltage to power the rest of the circuit. The equipment can then draw the current down to a regulated lower limit. Typically, before reaching that limit, the capacitor is recharged to the peak in the next half-cycle of the sine wave. This process is repeated over and over. The capacitor draws a pulse of current only during the peak of the wave. During the rest of the wave, the capacitor draws nothing.

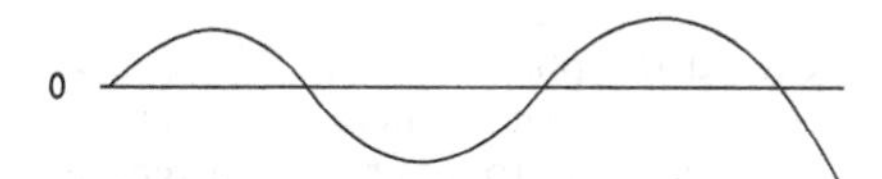

Fig. 7-26 60-cycle power.

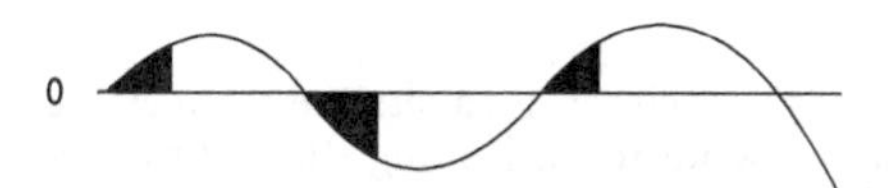

Fig. 7-27 Switched power supply current.

Fig. 7-28 Combination (distorted) waveform.

Symptoms of harmonics usually show up in the power distribution equipment that supports the nonlinear loads. There are two basic types of nonlinear loads: single-phase and 3-phase. Single-phase nonlinear loads are prevalent in offices, and 3-phase loads are widespread in industrial plants.

Neutrals

In a 3-phase, 4-wire system, neutral conductors can be severely affected by nonlinear loads connected to the 120-V branch circuits. Under normal conditions for a balanced linear load, the fundamental 60-Hz portion of the phase currents will cancel in the neutral conductor.

In a 4-wire system with single-phase nonlinear loads, certain odd-numbered harmonics called triplens (odd multiples of the third harmonic such as the 3rd, 9th, and 15th) don't cancel, but rather add together in the neutral conductor. In systems with many single-phase nonlinear loads, the neutral current can actually exceed the phase current. The danger here is excessive overheating because there is no circuit breaker in the neutral conductor to limit the current as there is in the phase conductors. Such excessive currents also cause high voltage drops between the neutral conductor and ground.

Circuit Breakers

Common *thermal-magnetic* circuit breakers use a bimetallic trip mechanism that responds to the heating effect of the circuit current. The bimetallic trip mechanism is designed to respond to the true-rms value of the current waveform and therefore will trip when it gets too hot. This type of breaker has a better chance of protecting against harmonic current overloads.

A *peak-sensing electronic trip* circuit breaker responds to the peak of current waveform. As a result, it won't always respond properly to harmonic currents. Since the peak of the harmonic current is usually higher than normal, this type of circuit breaker may trip prematurely at low current. If the peak is lower than normal, the breaker may fail to trip when it should.

The Harmonic Survey

A harmonic survey will give you a good idea whether you have a harmonic problem and where it is located. Here are a few guidelines to follow:

1. **Load inventory.** Make a walking tour of the facility and take a look at the types of equipment in use. If you see personal computers and printers, adjustable-speed motors, solid-state heater controls, and fluorescent fixtures with electronic ballasts, harmonics are likely.
2. **Transformer heat check.** Locate the transformers feeding those nonlinear loads and check for excessive heating. Also make sure the cooling vents are unobstructed.
3. **Transformer secondary current.** Use a true-rms meter to check transformer currents. You should measure and record the transformer secondary currents in each phase and in the neutral (if used). Then calculate the kVA delivered to the load and compare it to the nameplate rating. (Note, however, that if harmonic currents are present, the transformer can overheat even if the kVA delivered is less than the nameplate rating.) If the transformer secondary is a 4-wire system, compare the measured neutral current to the value predicted from the imbalance in the phase currents. (The neutral current is the vector sum of the phase currents and is normally zero if the phase currents are balanced in both amplitude and phase.) If the neutral current is unexpectedly high, triplen harmonics are likely and the transformer may have to be derated and its load reduced. Measure the frequency of the neutral current: 180 Hz would be a typical reading for a neutral current consisting of mostly 3rd harmonics.

4. **Sub-panel neutral current check.** Survey the subpanels that feed harmonic loads. Measure the current in each branch neutral and compare the measured value to the rated capacity for the wire size used. Check the neutral bus bar and feeder connections for heating or discoloration. An infrared temperature probe is useful for detecting excessive overheating on bus bars and connections.
5. **Receptacle neutral-to-ground voltage check.** Neutral overloading in receptacle branch circuits can sometimes be detected by measuring the neutral-to-ground voltage at the receptacle. Measure the voltage when the loads are on. Two volts or less is about normal. Higher voltages can indicate trouble, depending on the length of the run, quality of connections, and so on. Measure the frequency: 180 Hz would strongly suggest a presence of harmonics, and 60 Hz would suggest that the phases are out of balance.

Pay special attention to under-carpet wiring and modular office panels with integrated wiring that uses a neutral shared by 3-phase conductors. Because the typical loads in these two areas are computer and office machines, they are often trouble spots for overloaded neutrals.

Cures

The following are some suggestions for addressing some typical harmonics problems:

Overloaded neutrals

In a 3-phase, 4-wire system, the 60-Hz portion of the neutral current can be minimized by balancing the loads in each phase. The triplen harmonic neutral current can be reduced by adding harmonic filters at the load. If neither of these solutions is practical, install extra neutrals—ideally one for each phase. You can also install an oversized neutral shared by 3-phase conductors.

Derating transformers

One way to protect a transformer from harmonics is to limit the amount of load placed on it. This is called derating the transformer. For existing transformers, you will have to add new feeders, transformers, and branch circuit wiring—all good for business.

Test Equipment

The test equipment you use must be capable of taking both the true-rms phase current and the instantaneous peak phase current for each phase of the secondary.

8. GROUNDING

Grounding is critical for safety. Essentially, there are two purposes for grounding:

1. To provide a reliable return path for errant currents.
2. To provide protection from lightning.

At first, it would seem better *not* to provide a good return path for errant currents, thus making it harder for them to flow. Although this method would certainly be effective in reducing the size of *fault currents* (currents that flow where they are not intended), it would also allow them to flow more or less continually when they do occur.

Since fault currents pose the greatest danger to people, our primary concern is to eliminate these currents entirely. You do this by providing a clear path (one with virtually zero resistance—a "dead" short circuit) back to the power source so that these currents will be large enough to activate the fuse or circuit breaker. The overcurrent protective device will then cut off all current to the affected circuit, eliminating any danger. This also ensures that the circuit cannot be operated while the fault is present, which makes speedy repairs unavoidable.

You could say that you use grounding to make sure that when circuits fail, they fail all the way. Partially failed circuits are the dangerous ones because they can go unnoticed; therefore, grounding is essential.

The requirements for all types of grounding are covered in *Article 250* of the National Electrical Code. It is mandatory from both legal and safety standpoints that all systems be grounded according to these rules.

What Must Be Grounded

The following conductors must always be grounded:

1. One conductor of a single-phase, 2-wire system must be grounded.
2. The neutral conductor of a single-phase, 3-wire system must be grounded.
3. The center tap of a wye system that is common to all phases must be grounded.
4. A delta system must have one phase grounded.
5. In a delta system where the midpoint of one phase is grounded, the grounded conductor must be used as the neutral conductor.

Requirements for Various Systems

Two-wire dc systems that supply premises wiring must be grounded, unless one of the following situations exists:

1. The system supplies only industrial equipment in limited areas and has a ground detector.
2. The system operates at 50 volts (V) or less between conductors.
3. The system operates at more than 300 V between conductors.
4. The system is taken from a rectifier, and the ac supplying the rectifier is from a properly grounded system.
5. The system is a dc fire-protective signaling circuit that has a maximum current of 0.03 amperes.

Three-wire dc systems must have their neutral conductor grounded.

Alternating-current (ac) circuits operating at less than 50 V must be grounded in any of the following situations:

1. The circuit is installed as overhead wiring outside buildings.
2. The circuit is supplied by a transformer, and the transformer supply circuit is grounded.
3. The circuit is supplied by a transformer, and the transformer supply circuit operates at more than 150 V to ground.

AC systems operating at between 50 and 1000 V (these are the most common systems and the ones you are most likely to work on) and supplying premises wiring must be grounded if any one of the following conditions exists (see Fig. 8-1):

1. The system can be grounded so that the voltage to ground of ungrounded conductors cannot be more than 150 V.
2. The system is a 3-phase, 4-wire wye, and the neutral is used as a circuit conductor.
3. The system is a 3-phase, 4-wire delta, and the midpoint of one phase is used as a circuit conductor.
4. A grounded service conductor is not insulated.

The following systems are excepted from this requirement:

1. Systems used only to supply industrial electric furnaces used for melting and refining.
2. Separately derived systems used only to supply rectifiers that supply adjustable speed drives.
3. Other specialized systems (see exceptions in *Section 250.5(b)* of the NEC).

Circuits for cranes operating in Class 3 locations over combustible fibers may *not* be grounded.

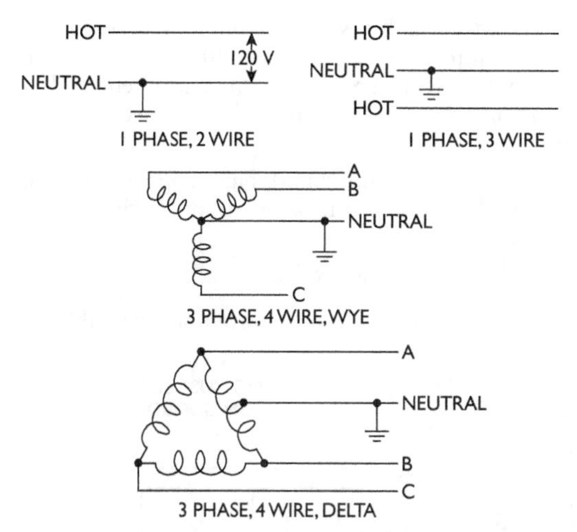

Fig. 8-1 Grounding connections for various systems.

The Grounding Electrode System

An electrical installation's grounding electrode system connects that system to ground. This is a critically important link in the grounding system and requires carefully chosen materials and methods.

All of the items listed below are suitable for grounding electrodes. All of the following (where available) must be bonded together, forming the grounding electrode system (see Figs. 8-2 through 8-5):

1. The nearest structural metal part of the structure that is grounded.
2. The nearest metal underground water pipe that is grounded.

3. An electrode, usually No. 4 (minimum) bare copper wire or reinforcing bar, at least 20 ft. long and in direct contact with the earth for its entire length. If a reinforcing bar is used, it must be at least ½ in. in diameter.
4. A ring of No. 2 (minimum) bare copper wire, at least 20 ft long and 2½ ft below grade, encircling the structure.
5. A ground rod, plate electrode, or made electrode.

Gas piping or aluminum electrodes are not acceptable.

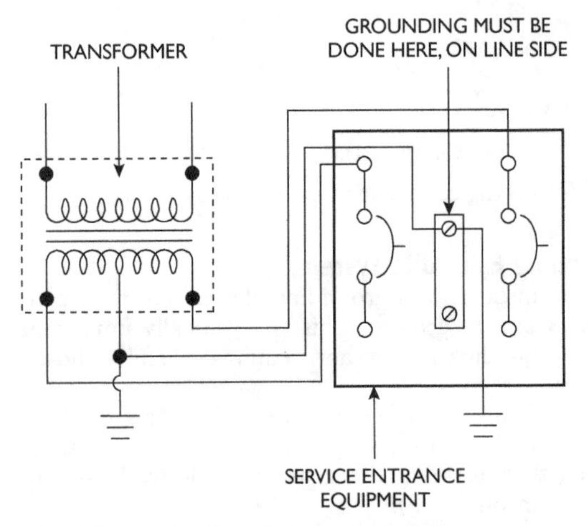

Fig. 8-2 Grounding of electric service.

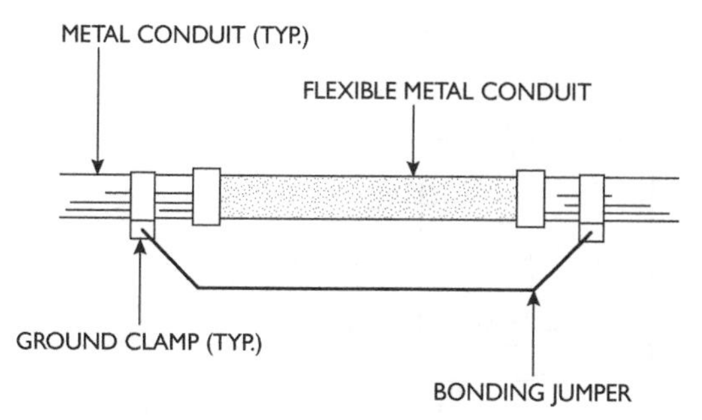

Fig. 8-3 Bonding around expansion joint.

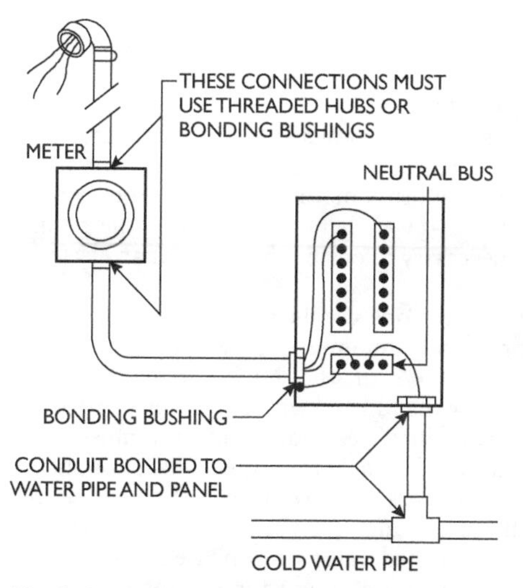

Fig. 8-4 Bonding conduit to water pipe.

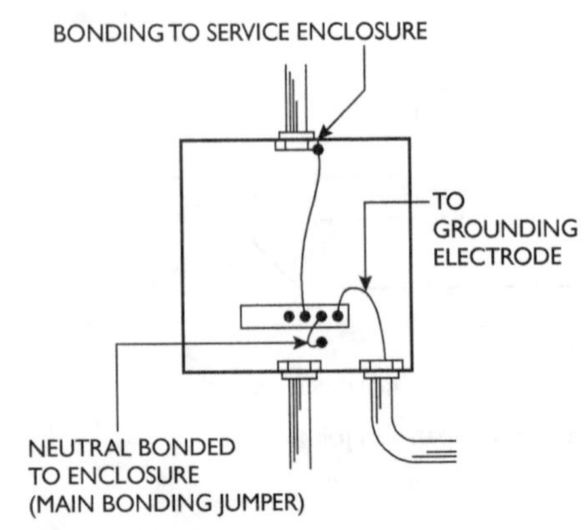

Fig. 8-5 Bonding enclosure.

Grounding Conductors

The grounding electrode conductor is the conductor that runs between the main service disconnect enclosure and a grounding electrode (most commonly a ground rod or cold-water pipe). This is a key element in completing the ground circuit, which is necessary to ensure the safety of an electrical system.

The grounding electrode conductor can be copper, aluminum, or copper-clad aluminum. Its size must be determined by *Table 250.66* of the NEC, or if the phase conductors are larger that 1100 kcmil copper or 1750 kcmil aluminum, the grounding electrode conductor must be at least 12½ percent of the area of the largest phase conductor. If the service conductors are paralleled, the size of the grounding electrode conductor must be based on the total cross-sectional area of the largest set of phase conductors.

The grounding electrode conductor cannot be spliced. (Bus bars and required taps are excepted.)

An equipment grounding conductor can be any of the following, and must be sized according to *Table 250.122* of the NEC:

- A copper (or other corrosion-resistant material) conductor
- Rigid metal conduits
- Intermediate metal conduits
- Electrical metallic tubing
- Listed flexible metal conduit, in lengths of 6 ft or less
- The armor of Type AC cables
- The sheath of mineral-insulated, metal-sheathed cables
- The sheath of Type MC cable
- Cable trays
- Cable buses
- Other metal raceways that are electrically continuous

Grounding electrode conductors can be installed in any of the following:

- Rigid metal conduit
- Intermediate metal conduit
- Rigid nonmetallic conduit
- Electrical metallic tubing
- Cable armor

Grounding electrode conductors that are No. 6 copper or larger can be attached directly to a building surface and are not required to be in a raceway, unless exposed to the possibility of physical damage.

Grounding Connections

The connection between a grounding electrode conductor and a grounding electrode must be accessible, permanent, and effective, because the safety of the entire system often depends on this connection.

Where metal piping systems are used as grounding electrodes, all insulated joints or parts that can be removed for service must have jumpers installed around them. The jumpers must be the same size as the grounding electrode conductor (see Fig. 8-6).

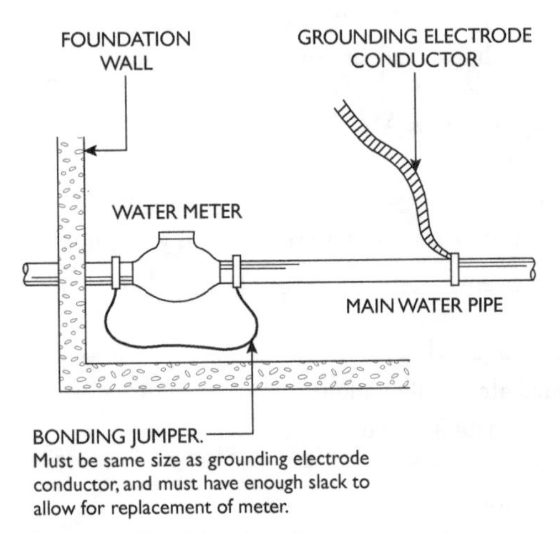

Fig. 8-6 Bonding around a water meter.

All grounding connections must be listed methods, and they must be permanent and secure.

If grounding connections are made in areas where they can be subjected to physical damage, they must be protected.

If more than one grounding conductor is present in a box or enclosure, all grounding conductors in the box must be connected. This connection must be made in such a way that the removal of any one device, for example, cannot affect the connection. Grounding connections cannot *feed through,* but must *pigtail* (see Fig. 8-7).

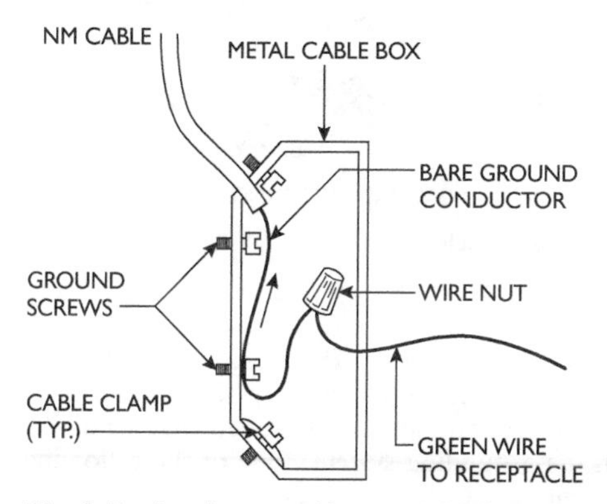

Fig. 8-7 Bonding metal box with NM cable.

Metal boxes must be connected to a grounding conductor, whether it be a conduit used as the grounding conductor or a separate grounding wire (see Fig. 8.8).

All paint or foreign substances must be removed from the area of grounding connections.

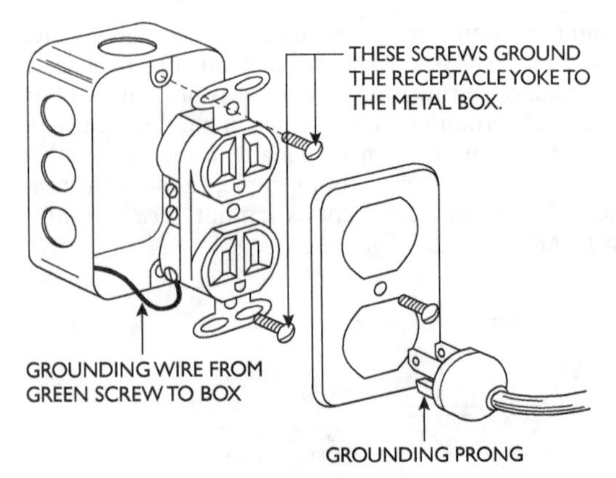

Fig. 8-8 Bonding receptacle.

Multiple Paths

If the use of multiple grounding paths creates a problem (as it can in certain circumstances, such as when a number of panels have a main bonding jumper installed that sends the circuit through the grounding system), one of the following steps can be taken:

1. Discontinue one or more of the grounding locations.
2. Change the locations of the connections.
3. Interrupt the conductor or path for the objectionable ground current.
4. Use other remedies approved by the authority having jurisdiction (the local inspector).

Steps 1–4 above can never be used as justification for not grounding every item connected to the system.

Direct current (dc) systems that are required to be grounded must have the grounding connection made at one or more supply stations. These connections cannot be made at individual services or on-premises wiring. (If the dc source is located on the premises, a grounding connection can be made at the first disconnecting means or overcurrent device.)

Connections for AC Systems

Alternating current (ac) systems that require grounding must have a grounding electrode conductor at each service that connects to a grounding electrode. The grounding electrode conductor must be connected to the grounded service conductor at an accessible location between the load end of the service and the grounding terminal, as shown in Fig. 8-9.

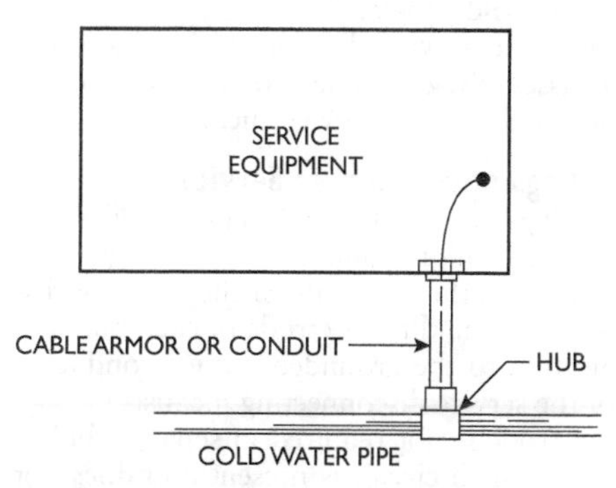

Fig. 8-9 Correct location of ground.

If the transformer supplying the load is located outside the building, a separate connection must be made between the grounded service conductor and a grounding electrode, either at the transformer or at another location outside the building.

No grounding electrode connection can be made on the load side of the service disconnecting means. (Grounding on the load side sometimes results in problematic and potentially dangerous ground-current loops.)

If an ac system that operates at less than 1000 V is grounded, the grounded conductor must be run to every service disconnecting means and bonded to every disconnecting means enclosure. This conductor must be run with the phase conductors and cannot be smaller than the grounding electrode conductor (as specified in *Table 250.94* of the NEC). If the phase conductors are larger than 1100 kcmil copper or 1750 kcmil aluminum, the grounded conductor must be at least 12½ percent of the area of the largest phase conductor. If the service conductors are paralleled, the size of the grounded conductor must be based on the total cross-sectional area of the largest set of phase conductors.

When more than one service disconnecting means are located in a listed assembly, only one grounded conductor must be run to and bonded to the service enclosure.

Two or More Buildings on a Common Service

If two or more buildings are supplied by a single service, the grounded system in each building must have its own grounding electrode connected to the enclosure of the building disconnecting means. The grounding electrode of each building must have a connection to the grounded service conductor on the load side of the service disconnecting means.

A grounding electrode is not required in separate buildings where only one branch circuit is present and does not require grounding.

Where two or more buildings are supplied by an ungrounded service, each structure must have a grounding electrode that is connected to the metal enclosure of the building or to the structure disconnecting means. The grounding electrode is not required in separate buildings where only one branch circuit that does not require grounding is present.

Disconnecting Means in a Different Building

When more than one building are under the same management and the disconnects are remotely located, the following conditions must be met:

1. The neutral is to be connected to the grounding electrode at the first building only.
2. Any building with two or more branch circuits has to have a grounding electrode (but only the first building should have its neutral connected to the grounding electrode). The equipment grounding conductor from the first building must be run to this second building with the phase conductors and be connected to the grounding electrode just mentioned.
3. The grounding conductor must be connected to the grounding electrode in a junction box located just inside or just outside the building.
4. A grounding conductor that is run underground must be insulated if livestock are present.

Grounding conductors must be sized according to *Table 250.122* of the NEC.

Separately Derived Systems

Separately derived systems are systems that are not taken (derived) from the normal source of power (supplied by the utility company). In almost all cases, the separate system that applies this power is a transformer. This transformer may be connected to the utility power on its primary side, but its secondary side is a separately derived system nonetheless and must be grounded as a new source of power, which in fact it is. When separately derived systems must be grounded (a few types don't have to be, mainly because they supply very limited power), the following requirements must be met:

1. Bonding jumpers must be used to connect the equipment grounding conductors of the derived system to the grounded conductor. The connection can be made anywhere between the service disconnecting means and the source, or it can be made at the source of the separately derived system if it has no overcurrent devices or disconnecting means. The bonding jumper cannot be smaller than the grounding electrode conductor (as specified in *Table 250.66* of the NEC). If the phase conductors are larger than 1100 kcmil copper or 1750 kcmil aluminum, the bonding jumper must be at least 12½ percent of the area of the largest phase conductor. If the service conductors are paralleled, the size of the bonding jumper must be based on the total cross-sectional area of the largest set of phase conductors.
2. A grounding electrode conductor must be used to connect the grounded conductor of the derived system to the grounding electrode. This connection can be made anywhere between the service disconnecting means and the source, or it can be made at the source of the separately derived system if it has no overcurrent devices or disconnecting means. The grounding electrode conductor must be sized as specified in *Table 250.94* of the NEC. If the phase conductors are larger than 1100 kcmil copper or 1750 kcmil aluminum, the bonding jumper must be at least 12½ percent of the area of the largest phase conductor. If the service conductors are paralleled, the size of the bonding jumper must be based on the total cross-sectional area of the largest set of phase conductors. (Class 1 circuits that are derived from transformers rated at no more than 1000 volt-amperes don't require a grounding electrode if the system's grounded conductor is bonded to the grounded transformer case.)

3. A grounding electrode must be installed next to (or as close as possible to) the service disconnecting means. The grounding electrode can be any of the following:
   a. The nearest structural metal part of the structure that is grounded.
   b. The nearest metal underground water pipe that is grounded.
   c. An electrode, usually No. 4 (minimum) bare copper wire or reinforcing bar, at least 20 ft long and in direct contact with the earth for its entire length. If a reinforcing bar is used, it must be at least ½ in. in diameter.
   d. A ring of No. 2 (minimum) bare copper wire, at least 20 ft. long and 2½ ft below grade, encircling the structure.
   e. A ground rod, plate electrode, or made electrode.

High-Impedance Ground Neutrals

Where high-impedance neutral systems are allowed (see *Section 250.36* of the NEC), the following requirements must be observed:

1. The grounding impedance must be installed between the grounding electrode and the neutral conductor.
2. The neutral conductor must be fully insulated.
3. The system neutral can have no other connection to ground, except through the impedance.
4. The neutral conductor between the system source and the grounding impedance can be run in a separate raceway.
5. The connection between the equipment grounding conductors and the grounding impedance (properly called the *equipment bonding* jumper) must be unspliced from the first system disconnecting means to the grounding side of the impedance.

6. The grounding electrode conductor can be connected to the grounded conductor at any point between the grounding impedance and the equipment grounding connections.

Enclosures

All metal enclosures must be grounded, except:

1. Metal boxes for conductors that are added to knob-and-tube, open wiring, or nonmetallic-sheathed cable systems that have no equipment-grounding conductor. The runs can be no longer than 25 ft and must be guarded against contact with any grounded materials.
2. Short runs of metal enclosures used to protect cable runs.
3. When used with certain Class 1, 2, 3, and fire-protective signaling circuits.

Equipment

All exposed noncurrent-carrying metal equipment parts that are likely to become energized must be grounded under any of the following conditions:

1. The equipment is within 8 ft vertically or 5 ft horizontally of ground or any grounded metal surface that can be contacted by persons.
2. The equipment is located in a damp or wet location, unless isolated.
3. The equipment is in contact with other metal.
4. The equipment is in a hazardous location.
5. The enclosure is supplied by a wiring method that provides an equipment ground (such as a metal raceway, metal-clad cable, or metal-sheathed cable).

6. When equipment is operated at over 150 V to ground, except:
   a. Nonservice switches or enclosures that are accessible only to qualified persons.
   b. Insulated electrical heater frames (by special permission only).
   c. Transformers (and other distribution equipment) mounted more than 8 ft above ground or grade.
   d. Listed double-insulated equipment.

The following types of equipment, regardless of voltage, *must* be grounded:

1. All switchboards, except insulated 2-wire dc switchboards.
2. Generator and motor frames of pipe organs, except when the generator is insulated from its motor and ground.
3. Motor frames.
4. Motor controller enclosures, except for ungrounded portable equipment or lined covers of snap switches.
5. Elevators and cranes.
6. Electrical equipment in garages, theaters, and motion picture studios, except pendant lamp holders operating at 150 V or less.
7. Electric signs.
8. Motion picture projection equipment.
9. Equipment supplied by Class 1, 2, 3, or fire-protective signaling circuits, except where specified otherwise.
10. Motor-operated water pumps.
11. The metal parts of cranes, elevators and elevator cars, mobile homes and recreational vehicles, and metal partitions around equipment with more than 100 V between conductors.

Unless specifically allowed to be otherwise, all cord-and-plug-connected equipment must be grounded.

Metal raceways, frames, and other noncurrent-carrying parts of electrical equipment must be kept at least 6 feet away from lightning protection conductors and lightning rods.

Methods

Equipment grounding connections must be made as explained below:

1. For grounded systems—by bonding the equipment grounding conductor and grounded service conductor to the grounding electrode conductor. (See Fig. 8-10.)
2. For ungrounded systems—by bonding the equipment grounding conductor to the grounding electrode conductor.

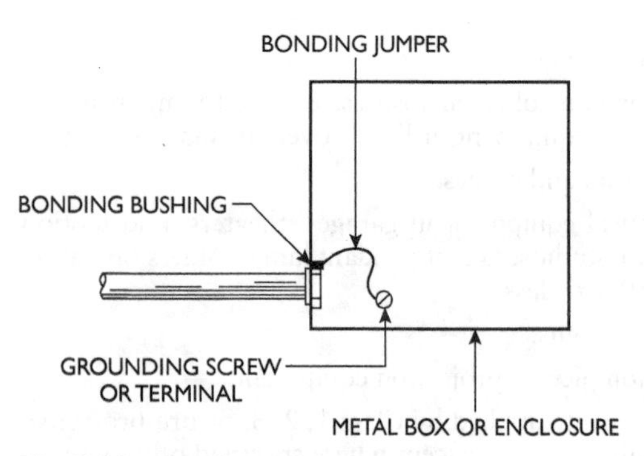

Fig. 8-10 Bonding service.

Grounding receptacles that replace ungrounded receptacles can be bonded to a grounded water pipe (see Fig. 8-11).

If there is no ground available in an existing box, it is also allowable to replace a nongrounding type of receptacle with a GFI receptacle. Care must be taken, however, that the grounding conductor from the GFI receptacle is not connected to a ground conductor on the load side of the GFI. That would jeopardize the safety of the installation.

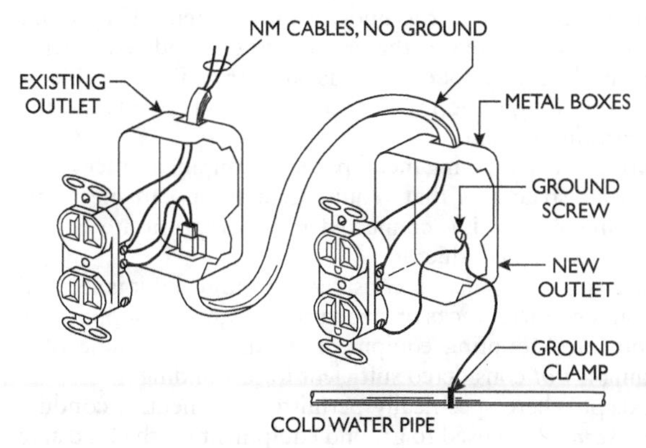

Fig. 8-11 Method of adding a grounded receptacle to an ungrounded source.

The conductor path to ground from equipment and metal enclosures must:

1. Be permanent and continuous.
2. Have enough capacity for any fault current imposed on it.
3. Have a low enough impedance so as not to limit the voltage to ground.

The earth cannot be used as the only grounding conductor. (In other words, you cannot ground something by simply attaching a wire to its case and pushing the wire into the ground.)

Only one grounding electrode is allowed at each building. Two or more grounding electrodes that are bonded together are considered to be the same as one grounding electrode.

Metal sheaths of underground service cables can be considered grounded only because of their contact with the earth and bonding to the underground system. They don't need to be connected to the grounding electrode conductor or grounding electrode. That is also true if the cable is installed underground in metal conduit and is bonded to the underground system.

Noncurrent-carrying metal parts of equipment, raceways, and other structures that require grounding can meet this requirement by being connected to an appropriate equipment grounding conductor.

Electrical equipment is considered grounded if it is secured to grounded metal racks or structures designed to support the equipment. Mounting equipment on the metal frame of a building is *not* considered sufficient for grounding.

Except where specifically permitted, the neutral conductor must *never* be used to ground equipment on the load side of the service disconnect.

If a piece of equipment is connected to more than one electrical system, it must have an appropriate ground connection for each system.

Bonding

Bonding is merely the connecting of metal parts to form a complete grounding system. Bonding is essential for maintaining the continuity of the grounding system; it is also important protection against voltage surges caused by lightning or other fault currents.

The following parts of service equipment must be bonded together:

1. Raceways, cable trays, cable armor, or sheaths.
2. Enclosures, meter fittings, etc.
3. A raceway or armor enclosing a grounding electrode conductor.

One exposed method must be provided for the bonding of other systems (such as telephone or cable TV systems). This can be done by one of the following:

1. A grounded metal raceway (must be exposed).
2. An exposed grounding electrode conductor.
3. Any other approved means, such as extending a separate grounding conductor from the service enclosure to a terminal strip.

The various components of service equipment must be bonded together by one of the following methods:

1. Bonding the components to the grounded service conductor.
2. Making the connection with rigid or intermediate metal conduit, made up with threaded couplings or tight threadless couplings and connectors.. Standard locknut and bushing connections are *not* sufficient.
3. Connecting bonding jumpers between the various items.
4. Using grounding locknuts or bushings, with a conductor bonding the various items.

Metal raceways and metal-sheathed cables that contain circuits operating at more than 250 V to ground must be bonded in the same way as service equipment, as outlined above. (The exception is a connection to the grounded service conductor.) If knockouts in enclosures or boxes are field-cut, standard methods can be used.

If the uninsulated neutral of a metal service cable is in direct contact with the metal armor or sheath, the sheath or armor is considered grounded without any further connection.

A receptacle's grounding terminal must be bonded to a metal box by one of the following methods (see Fig. 8-12):

1. By connection with a jumper wire (commonly called a *pigtail*).
2. By receptacle yokes and screws approved for this purpose.
3. By direct metal-to-metal contact between the box and a receptacle yoke. (This applies to surface-mounted boxes *only.*)
4. By installation in floor boxes listed for this purpose.
5. By connection to an isolated grounding system, where eliminating electrical noise in sensitive circuits is required.

All electrical components that are allowed to act as equipment conductors must be fitted together well in order to ensure electrical continuity. They *must* be well-bonded. These systems include metal raceways, cable trays, cable armor, cable sheaths, enclosures, frames, and fittings.

All metal raceways must be made electrically continuous. This requires special care at expansion joints and other interruptions.

Main bonding jumpers connect the grounded service conductor and the service enclosure. This connection should be made according to the instructions supplied by the manufacturer of the service equipment. The bonding jumper cannot be smaller than the grounding electrode conductor (as specified in *Table 250.66* of the NEC). If the phase conductors are larger than 1100 kcmil copper or 1750 kcmil aluminum, the bonding jumper must be at least 12½ percent of the area of the largest phase conductor. If the service conductors are paralleled, the size of the bonding jumper must be based on the total cross-sectional area of the largest set of phase conductors.

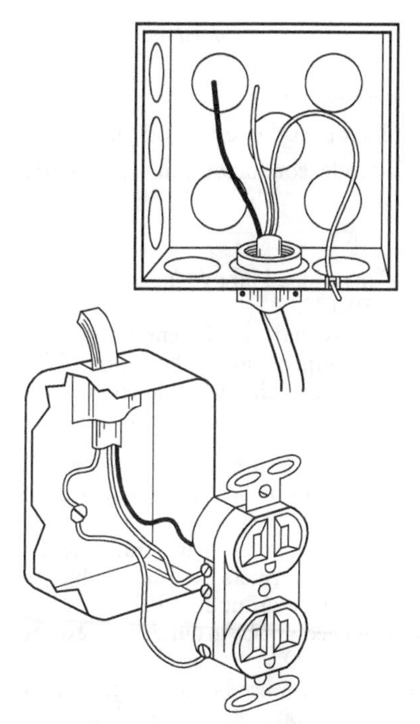

Fig. 8-12 Box bonded with clip (top); box bonded with screw connection (bottom).

An equipment bonding jumper on the load side of the service must be sized according to *Table 250.122* of the NEC, based on the largest overcurrent that protects the conductors in the raceways or enclosures.

The bonding jumper mentioned above can be inside or outside the equipment being bonded. If installed outside, it can be no more than 6 feet long and must be routed along with the equipment.

Interior metal water piping systems *must* be bonded to one of the following:

1. The service equipment enclosure.
2. The grounded conductor (at the service only).
3. The grounding electrode conductor (unless it is too small).
4. The grounding electrode.

The bonding jumper mentioned above must be sized according to NEC *Table 250.122.*

Other metal piping systems that may be energized must be bonded. The size of the jumper must be based on *Table 250.122,* according to the rating of the circuit likely to energize the piping.

Other Requirements

All instruments and equipment in or on switchboards must be grounded.

When the primary voltage of instrument transformers exceeds 300 V, and the transformers are accessible to unqualified persons, they must be grounded. If inaccessible to unqualified persons, they need not be grounded unless the primary voltage is more than 1000 V.

Instrument transformer cases must be grounded if they are accessible to unqualified persons.

High-voltage systems that require grounding must meet the same requirements for grounding as low-voltage systems. (For exceptions, see *Section 250.184* of the NEC.)

Surge Arresters

Surge arresters are used to send surge voltages (unusually large voltages such as may be caused by a nearby lightning strike) directly to ground, rather than allowing them to affect interior wiring. They are extremely important under some circumstances. (See Fig. 8-13.)

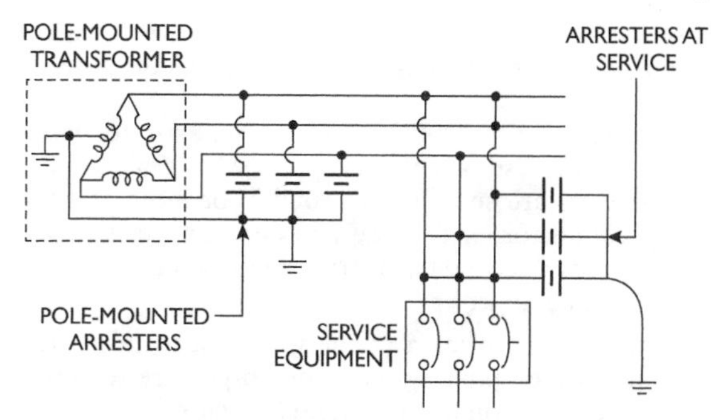

Fig. 8-13 Connection of surge arresters.

Any surge arresters installed must be connected to all ungrounded conductors.

The rating of a surge arrester for circuits *under* 1000 V must be equal to or greater than the operating voltage of the circuit it protects.

The rating of a surge arrester for circuits *over* 1000 V must be at least 125 percent of the operating voltage of the circuit it protects.

Surge arresters can be installed indoors or outdoors. Unless specifically approved for such installation, they must be located where they are not accessible to unqualified persons.

Conductors connecting the surge arrester to the system being protected must be as short as possible.

All surge arresters must have a connection to one of the following:

1. The equipment grounding terminal.
2. The grounded service conductor.
3. The grounding electrode conductor.
4. The grounding electrode.

For services of 1000 V or less, the minimum size for wires connecting surge arresters is No. 14 copper or No. 12 aluminum.

For services of over 1000 V, the minimum size for wires connecting surge arresters is No. 6 copper or No. 6 aluminum.

When circuits are supplied by 1000 V or more and the grounding conductor of a surge arrester protects a transformer that feeds a secondary distribution system, one of the following must be done:

1. If the secondary has a connection to a metal underground water pipe system, the surge arrester must have a connection to the secondary neutral.
2. If the secondary does not have a connection to a metal underground water pipe system, the surge arrester must be connected to the secondary neutral through a spark gap device. (See *Section 280.24(B)* of the NEC for spark gap device specifications.)

Special Permission

In extraordinary cases, an *authority having jurisdiction* (usually the local electrical inspector) can give special permission for grounding connections that are not specified in the NEC. That is done only under very unusual conditions, however.

9. CONTACTORS AND RELAYS

There are many different types of electrical controls that can be used to build complex electrical circuits. They range from tiny solid-state switches to the gigantic switches that the power companies use to take their generators on or off line.

For most commercial- and factory-type applications, however, the common types of control equipment are the medium-sized and more basic types, which are far less expensive and easier to locate. Most of these items can be purchased through a local electrical supply wholesaler or through commercial suppliers, such as W. W. Grainger and McMaster-Carr Supply.

Relays

Relays are probably the most important and most common of all electrical controls. A relay uses a control current from one source to control a separate circuit by using electrical, magnetic, and mechanical principles. A diagram of a very basic relay is shown in Fig. 9-1. The circuit that is controlling the relay is connected to terminals 1 and 2, and the circuit being controlled is connected to terminals 3 and 4. When the controlling circuit is energized, a current will flow through the magnetic coil, which is attached to terminals 1 and 2; this will cause the *solenoid* (defined below) to pull closed and so close the switch, which will complete the circuit that is being controlled.

The solenoid is nothing more than a coil of small copper wire with an opening in its middle into which the iron bar slides in and out. There is a small spring on the bottom of the iron bar to keep it pushed up (keeping the switch open) at all times, except when the solenoid is energized. When the solenoid is energized, it becomes an electromagnet and pulls the iron into its middle.

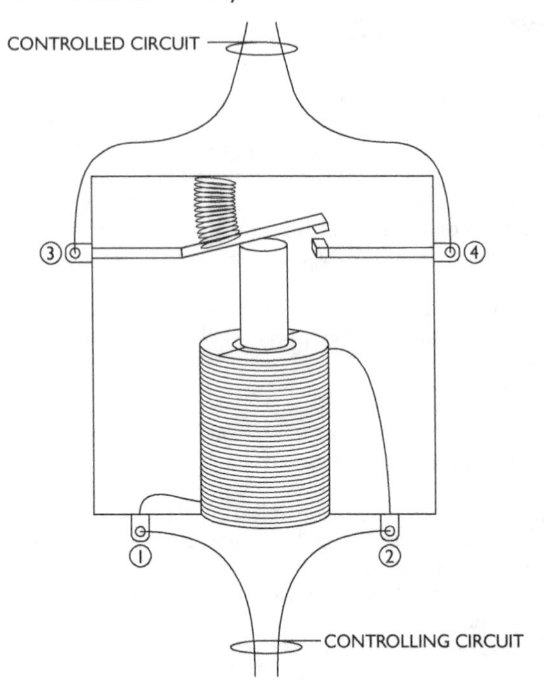

Fig. 9-1 Basic relay.

The most common type of control relay is shown in Fig. 9-2. This type of relay is also called an *eight-pin relay*, because it has eight terminals. The terminals are in the form of metal pins that fit into the relay's base, where wire can be connected. This type of relay is very small in size, standing only about 3 in. tall including the case, and is common and inexpensive. It can be found at electrical supply houses, commercial supply houses, electronic supply houses, and even at some hardware stores. The circuitry for this type of relay is shown in Fig. 9-3, along with a description of each of the symbols shown in the diagram. The power to the relay coil (the controlling circuit) connects to terminals 2 and 7. Terminals 1 and 8 are what we can call *common* terminals (because they are common to two separate sets of contacts), 6 and 3 are *normally open* contacts, and 4 and 5 are *normally closed* contacts.

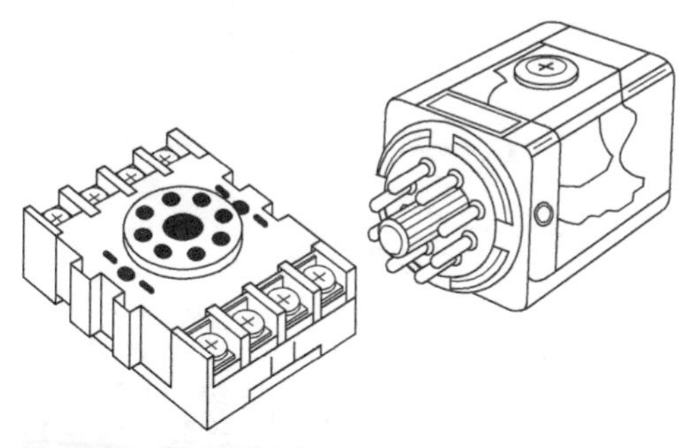

Fig. 9-2 Control relay.

The operation of the relay, shown by using contacts 1, 3, and 4, is as follows: When power is sent to the relay coil (terminals 2 and 7), the coil energizes, and the various contacts in the relay either open or close. Contact 3 is not touching contact 1 before the power comes to the coil (which is why it is called a normally open contact); but when the coil energizes, the contacts close, completing a circuit with terminals 1 and 3. Before the coil is energized, contacts 1 and 4 are touching (which is why they are called normally closed contacts); however, when the coil is energized, these contacts come apart (called opening), breaking the flow of any current flowing through the connection.

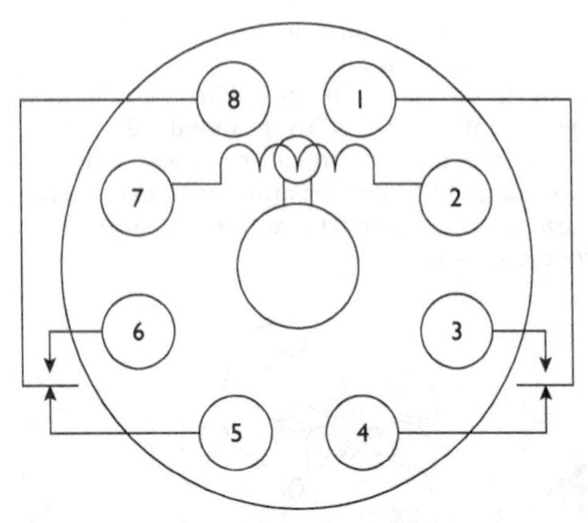

Fig. 9-3 Eight-pin relay.

Fig. 9-4 shows this relay in use in a circuit. The relay is controlled according to a certain 50-horsepower motor in a factory. When the motor is turned off, the relay coil stays deenergized; it becomes energized when the motor is turned on by its starter. When the motor turns on, the coil will become energized, and contacts 1 and 3 will close. This action causes a pilot light to turn on, which shows the factory's operations manager (in a remote location in the factory) that this particular motor has been turned on and is running. Contacts 5 and 8 are touching before power comes to this relay's coil, but when the relay is energized the contacts open, breaking the circuit going through these contacts. In this instance, when the contacts open, they break a control circuit that controls another motor in the factory. For safety reasons, this motor must be turned off when the first motor is operating. When contacts 5 and 8 break, the control circuit for the second motor is broken, shutting the motor off.

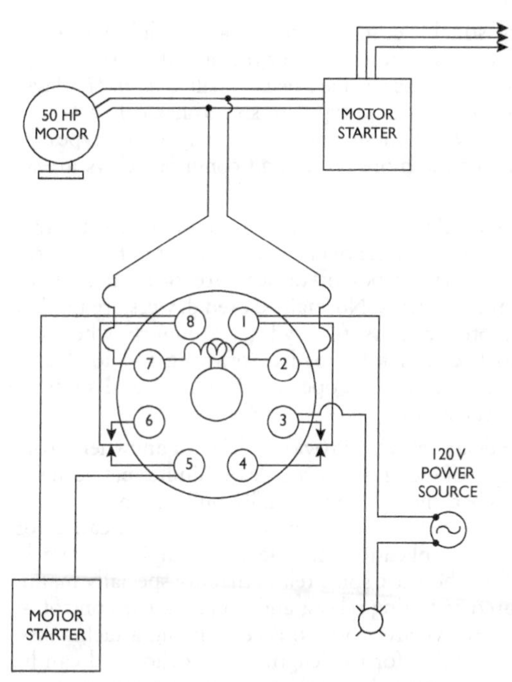

Fig. 9-4 Eight-pin relay in circuit.

The eight-pin relay in Fig. 9-4 is typical for all relays, which come in many sizes and types depending upon the application. In selecting relays, care must be taken that the voltages and currents at which they are used never exceed their rated levels. This type can handle heavier currents and greater voltages than the eight-pin type. When installing any relays that use household voltages, care must be taken to install the relays in some type of fireproof box or enclosure. Such boxes are available at any electrical wholesale house for a very reasonable cost. Be sure that any such box is listed with Underwriters Laboratories for the intended use.

The relays just described are usually called *control* relays. Larger relays of the same type are sometimes called *power* relays. But, beyond these, there are many other types of relays. Some of the more useful and common relays are as follows:

Current relays. These are specially made to trip when a certain level of current is present in the control circuit. These types of devices are often used with hydraulic systems. Normally closed devices are used in the motor circuits for hydraulic pumps; when the pump becomes fully loaded (drawing full load current), the relay trips, opening the circuit and shutting down the motor.

Time delay relays. Provide a delay of anywhere from a couple of seconds to a couple of minutes between the time when the solenoid actually pulls in and the time when the contacts open or close. While not called for in every application, this feature can be extremely helpful. There are some relays that are specially manufactured as timing relays; and some regular control or power relays are made to accept timing attachments, are adjustable for the length of the delay, and can be purchased at an electrical or electronic supplier.

Momentary contact relays. The relays covered so far have been *maintained* contact relays. That is, once the relay is tripped, the contacts make or break contact and maintain that condition. *Momentary* contact relays are different in that, once they trip, they make or break contact for a moment and then return to the previous condition. They are especially effective when used with other control mechanisms, such as motor starters, and with control systems that require a momentary burst of energy instead of a constant one.

Mechanically held relays. The standard type of relay is *electrically* held. That is, once the solenoid is pulled into position, it is held there by a continued electrical current. If the current were to stop, the solenoid would return to its original position. Mechanically held relays have a built-in mechanism that locks the solenoid into place once the relay is tripped. The electrical power can then be discontinued without affecting the position of the relay. This is often a desirable characteristic because it eliminates noise and waste of energy. Note that this type of relay needs a burst of power to turn on and a burst of power to turn off. Normal relays need power to turn on, then turn off when power is absent.

Alternating relays. These are relays that automatically switch power between two different circuits. They are commonly used for duplex pump circuits and similar systems. When power is called for by the control circuit (activated by various sensors), power is delivered to the first pump (or other device) until the control circuit shuts it back off. Then, when power is again called for, it is sent to the second pump. The alternating relay (also called simply an *alternator*) will switch power from one to the other indefinitely.

Maintenance

Since relays contain a number of moving parts, performing routine maintenance on them is crucial to continuing good operation.

Contacts

Once a relay is properly adjusted, all that is required for maintenance are certain periodic checks and cleaning. It is not necessary to readjust the relay unless the checks indicate that readjustment is needed. In almost all cases, poor operation of relays is caused by problems with the contacts (dirt or oxidation), not by maladjustment.

Under normal operating conditions, most contacts are self-cleaning. That is, they tend to wipe themselves clean with every activation. However, under certain conditions, dirt or oxidation can build up on the contacts nonetheless, particularly on relays that operate only occasionally. When this occurs, a contact burnishing tool should be used to clean the contact points.

A burnishing tool is made of very fine abrasive material and is designed to remove only foreign matter from contact points. Files should never be used for this work in place of a burnishing tool, since even moderate use of a file would remove plating from a contact along with the dirt or oxidation.

When using a burnishing tool, it is important that no more than normal contact pressure be applied to the contact being burnished. That is, the relay should be operated by hand, but no extra pressure should be applied at the points themselves with tools or fingers. A gentle wiping action should be applied between the contacts until all the foreign matter is removed.

Note: Solvents (such as carbon tetrachloride or other similar solvents) should not be used because they leave a film or residue on the contacts.

Additional checks to be made for routine maintenance (see Fig. 9-5) include the following:

1. The residual screw should be checked by visual inspection. There should be visual clearance (sighted against a light and held in the energized position) all the way across the heel piece. During this check, the armature motion is observed to see that it operates freely and doesn't bind at the pin or yoke.

2. The contact assembly should be checked visually also. When the armature is operated by hand, there should be a slight but perceptible deflection of the stationary springs.

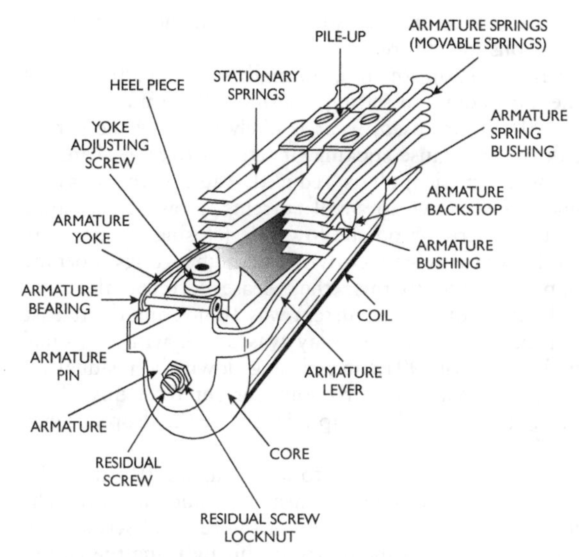

Fig. 9-5 Relay detail.

Adjustments

This routine maintenance should be done before any attempt is made to adjust the relay. If the relay doesn't meet these requirements, it can be readjusted as described in the following instructions.

Although there are many different types of relays, the basic principles of operation and adjustment, in general, will apply to all types. The adjustments described in the following paragraphs cover the operations of a common type of relay used in many applications on communications equipment.

The actual adjustment of the relay can be broken down into two sections: adjustment of the armature assembly, and adjustment of the contact assembly "pile-up." It is important that no attempt be made to adjust a relay without having the specifications for that relay.

The residual adjustment is normally set from .001 to .004 in. The most common setting is around .0015 in. If settings below .001 in. are used, wear is likely to allow the gap to close down and cause sticking. If settings over .004 in. are used, the magnetic pull is reduced, causing the relay to become less sensitive to small currents. However, in some applications, settings on the low side (closing the gap) are used to lower the release current, and high settings (opening the gap) are used to raise the release current value. For example, if the operating current was 10 mA and the release or drop-out current for a relay was 5 mA with a normal residual setting of .0015 in., closing down the residual air gap might decrease the drop-out current to 3 mA. Also, increasing the residual air gap might raise the drop-out current to 7 mA.

To set the residual air gap to .0015 in., a .0015-in. feeler gauge is inserted between the armature and the core. The gauge must not prevent the end of the residual screw from striking the core. It is handy to do this by using the end of the gauge with the hole in it and positioning it so that the residual screw passes through it. See Fig. 9-6.

The relay is then held by hand in the energized position. The gauge should then be a snug fit, but not binding. If the gauge is loose or binds, the residual screw lock nut may be loosened with a box wrench or an open-end wrench and the residual screw adjusted for proper clearance. Be sure to check the clearance after tightening the lock nut.

The air line may be set from .001 to .008 in. Generally a setting of .003 to .004 in. is used. The proper feeler gauge is inserted between the heel piece and the armature; be careful not to insert it far enough to come between the armature and the core. See Fig. 9-7. The yoke adjusting screw is slightly loosened with an offset screwdriver. The armature is then held against the core and heel piece. Tap the armature lightly to make a snug fit on the gauge. Be sure to check both sides of the gap—it should be even all the way across. The yoke adjusting screw is then tightened.

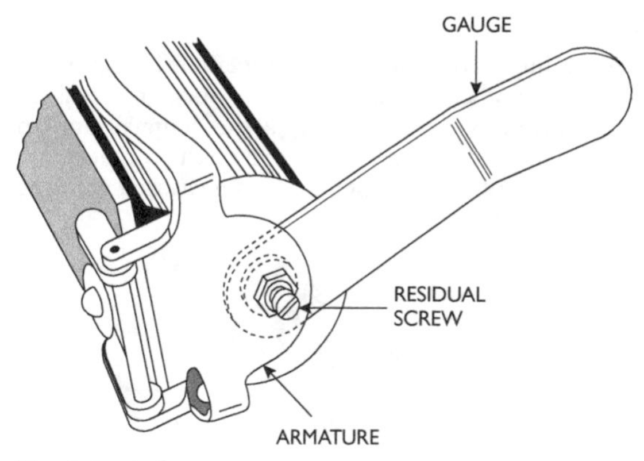

Fig. 9-6 Relay detail.

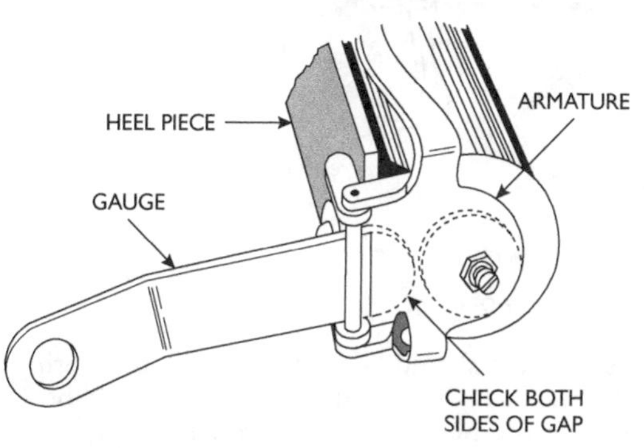

Fig. 9-7 Relay detail.

The armature end play should be checked. This can be done by inserting a feeler gauge between the armature bearing and the yoke. The clearance should be between .010 and .030 in. The armature must also move freely without any binding.

Tension

More relay failures are caused by improper tensioning of the springs than any other cause. Thus it is important to tension the springs properly before adjusting the contacts and to recheck the tension after the adjustments to the contacts. See Fig. 9-8. Tension of the break contacts is measured with the relay in the deenergized position.

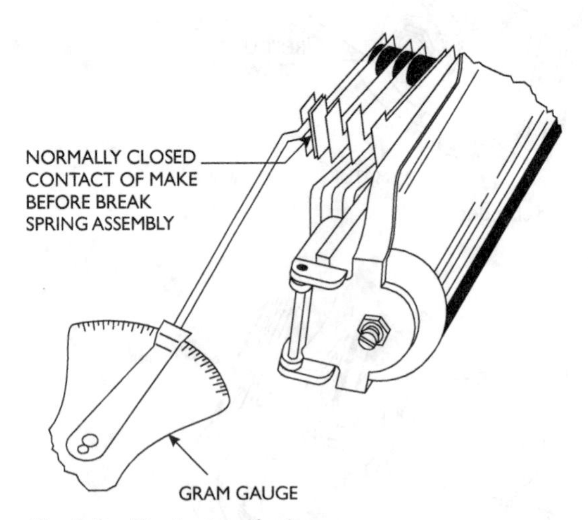

Fig. 9-8 Tensioning of relay.

The measurement should be made with all other springs lifted off the one being measured. The tension should be 25 to 35 grams at the point where the contacts just break. Tension on the make contacts is taken care of by the proper adjustment of the springs (gauging).

To increase the spring tension, a spring-bending tool is placed on the armature spring back near the fillet. Next the bender is turned slightly toward the heel piece and is drawn along the spring slowly to the armature bushing. See Fig. 9-9. This should leave a slight box (no more than 1/32 in.) in the spring. The concave (depressed) side of the bow will be facing the heel piece. Again, the spring bender is applied near the fillet, and the spring is bent toward the heel piece until the bow is flattened out. This should give the proper tension to the spring. To reduce the spring tension, the spring is bowed slightly in the opposite direction.

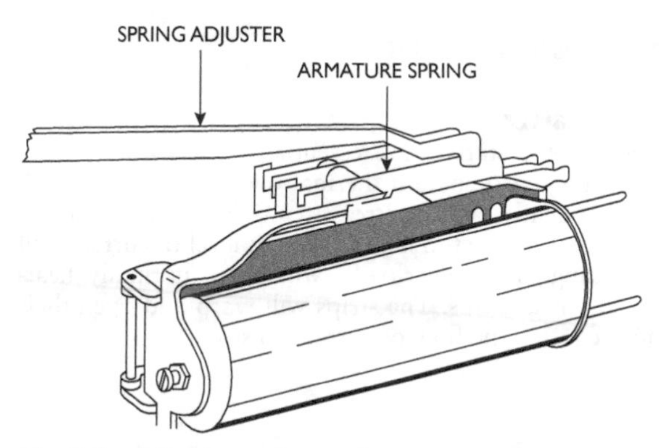

Fig. 9-9 Adjusting spring tension.

Contactors

One of the more commonly used types of relays is the lighting contactor. A typical unit is shown in Fig. 9-10. Contactors are used to turn on multiple circuits (typically lighting circuits) simultaneously. Contactors come in both electrically and mechanically held varieties and may control between 2 and 12 circuits (2-pole through 12-pole contactors). Circuit ampacities are typically between 20 and 40 amps.

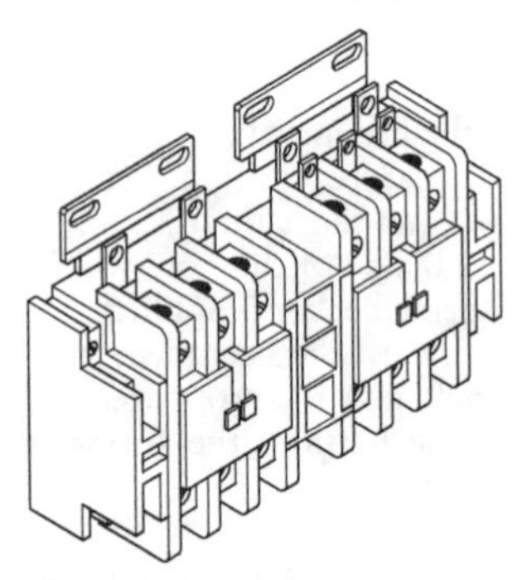

Fig. 9-10 Lighting contactor.

Magnetic Starters

Magnetic motor starters are essentially heavy-duty relays. However, they also contain thermal overload relays, which can independently open control circuits or power circuits if the current levels rise too high. The high level of current will cause heating in the overloads, which are normally heat-sensitive bimetal strips. The strips will warp and open their contacts, causing the flow of current to stop.

10. WELDING

Welding is important for electrical installations in industrial situations. In many cases, runs of conduit and pieces of equipment must be welded into place, rather than being secured with strap or clamps, as is common in other electrical installations.

Because of the development of new welding processes and new types of metals that can be welded, welding has become the most important metal joining process by far. Soldering and brazing have also been used to join metal, but generally both (especially soldering) have characteristics far less desirable than welding, and they are seldom used today.

Welding Processes

The commonly used welding processes may be grouped into four general categories:

1. Gas oxyacetylene welding.
2. Shielded metal-arc welding. (Shielding connotes the creation of an environment of controlled gas or gases around the weld zone to protect the molten weld metal from contamination by oxygen and nitrogen in the atmosphere.)
3. Gas metal-arc welding.
4. Gas tungsten-arc welding.

Oxyacetylene Welding

The oxyacetylene gas welding process is shown in Fig. 10-1. The equipment required for oxyacetylene welding is shown in Fig. 10-2. This welding process has been in use for more than 100 years, with the methods essentially the same as when the process originated. The process is extremely flexible and is one of the most inexpensive as far as equipment is concerned. It is used most for maintenance welding, small-pipe welding, auto body repairs, welding of thin materials, and sculpture work. The high temperature generated by the equipment can also be used for soldering, hard soldering, and brazing.

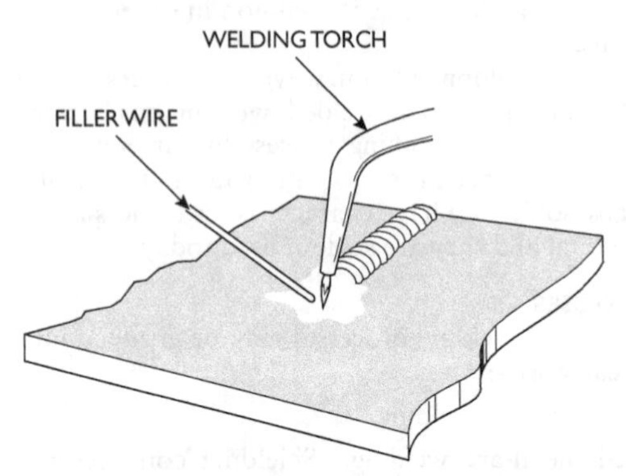

Fig. 10-1 The oxyacetylene gas welding process.

A variation of the process allows for flame cutting with a special type of torch. Bringing the metal to a high temperature and then introducing a jet of oxygen burns the metal apart. This torch is a primary cutting tool for steel.

The oxyacetylene gas welding process fuses metals by heating with a gas flame obtained from the combustion of acetylene with oxygen. Welding can be done with or without filler metal. An oxyacetylene flame is one of the hottest flames produced by portable equipment—as high as 6300°F.

The flame melts the two edges of the metal pieces that are being welded. It also melts any filler metal that is added to fill the gaps or grooves, so the molten metal can mix rapidly and smoothly. The acetylene and the oxygen gases flow from separate cylinders to the welding torch, where they are mixed and burned at the torch tip.

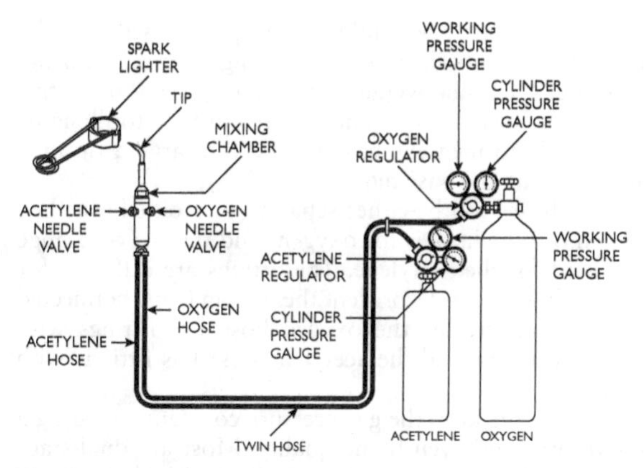

Fig. 10-2 Oxyacetylene welding equipment.

Torches and Flames

The proportions of oxygen and acetylene determine the type of flame. The three basic types of flame mixtures are called *neutral, carburizing,* and *oxidizing.*

The neutral flame is generally preferred for welding. It has a clear, well-defined white cone indicating the best mixture of gases, and no gas is wasted.

The carburizing flame uses more acetylene, has a white cone with a feathery edge, and adds carbon to the weld.

The oxidizing flame, using more oxygen, has a shorter envelope and a small, pointed white cone. This flame oxidizes the weld metal and is used only with specific metals.

Flame cutting is accomplished by adding an extra oxygen jet to burn a cut into the metal.

The standard torch can be a *combination type* used for welding, cutting, and brazing. The gases are mixed within the torch. A thumb-screw needle valve controls the quantity of gas flowing into a mixing chamber. A lever-type valve controls the oxygen flow for cutting with a cutting torch or attachment. Various types and sizes of tips are used with the torch for specific applications of welding, cutting, brazing, or soldering. The usual welding outfit has three or more tips to fit a variety of jobs. Too small a tip will take too long or will be unable to melt the base metal. Too large a tip may result in burning the base metal.

The gas hoses can be either separate or molded together. The green or blue hose is for oxygen, and the red or orange hose is used for the acetylene. The fittings are different for each of the hoses so as to prevent them from being connected incorrectly. Specifically, the oxygen hose has fittings with right-hand threads, and the acetylene hose has fittings with left-hand threads.

Gas regulators keep the gas pressure constant, ensuring a steady volume and even flame quality. Most are dual-stage regulators and have two gauges, one displaying the pressure in the cylinder and the other showing the pressure entering the hose. Gases welding uses oxygen and acetylene primarily, but other gases (such as hydrogen, city gas, natural gas, propane, or mapp gas) are sometimes used instead of acetylene. These other gases are used only for specific applications. Acetylene is the preferred gas in most instances because it has a higher burning temperature. It produces more and better welds per given period of time.

The *gas cylinders* for acetylene contain porous material saturated with acetone. Since acetylene can't be compressed with safety to more than 15 psi, it is dissolved in the acetone. This keeps it stable and allows for pressures of up to 250 psi. Acetylene cylinders should always stand upright.

Oxygen cylinder capacities vary from 60 to 300 cu. ft., with pressures up to 2400 psi. The maximum charging pressure is always stamped on the cylinder and should never be exceeded.

Shielded Metal-Arc Welding

This is perhaps the most popular welding process in use today. The high quality of the weld produced by the shielded arc process, plus the high rate of production, has made it a replacement for many other fastening methods. The process can be used in all positions and will weld a wide variety of metals. The most popular use, however, is the welding of mild carbon steels and low-alloy steels.

Shielded metal-arc welding is an arc welding process that produces coalescence between a covered metal electrode and the work by heating with an arc. Shielding is obtained by the decomposition of the electrode covering. The electrode also supplies the filler metal. Figure 10-3 shows the covered electrode, the core wire, the arc area, the shielding atmosphere, the weld, and the solidified slag.

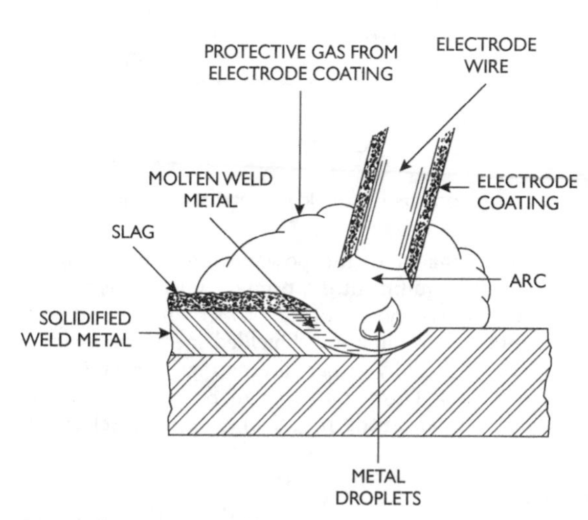

Fig. 10-3 Shielded metal-arc welding.

This manually controlled process welds all *nonferrous* (non-iron, such as copper or bronze) metals ranging in thickness from 18 gauge to the maximum encountered. For material with a thickness of ¼ in. or more, a beveled edge preparation is used with the multipass welding technique. The process allows for all-position welding with the arc visible and under the welder's control. Slag removal is required. The major components required for shielded metal-arc welding are shown in Fig. 10-4.

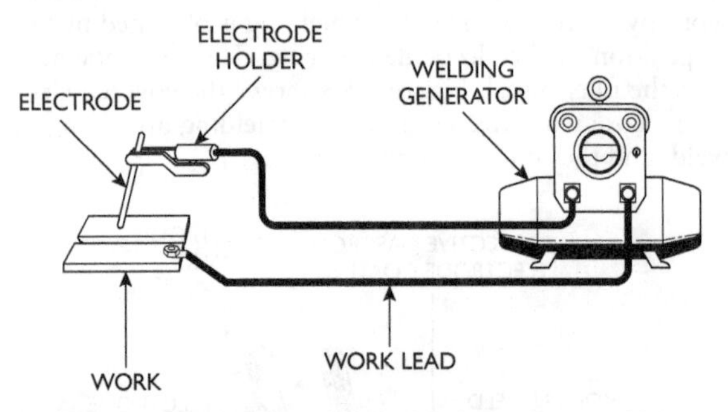

Fig. 10-4 Components for shielded metal-arc welding.

The welding machine (the power source) is the major piece of required equipment. Its primary purpose is to provide sufficient electric power of the proper current and voltage to maintain a welding arc. Shielded metal-arc welding can be accomplished by either alternating current (ac) or direct current (dc). Direct current can be employed either *straight* (with the electrode negative) or *reverse* (electrode positive).

Types of Welders

A variety of welding machines are used, each having its specific advantages or special features.

- The *ac transformer* type is simple, inexpensive, and quiet.
- The *transformer-rectifier* machine converts ac power to dc power and provides direct current at the arc.
- The *ac-dc transformer-rectifier* welding machine combines the features of both the transformer and the rectifier.
- The *direct current generator* is probably the most versatile welding power source.
- The *dual-control single-operator generator* is a conventional type that allows the adjustment of the open-circuit voltage and the welding current. When electric power is available, the generator is driven by an electric motor. Away from power lines, the generator can be driven by a gasoline internal combustion engine or a diesel engine.

Operating the Welder

The operator holds the electrode holder, which firmly grips the electrode and carries the welding current to it. The insulated pincer-type holders are the most popular. Electrode holders come in various sizes and are designated by their current-carrying capacity.

The welding circuit consists of the welding cables and connectors that provide the electrical circuit conducting the welding current from the machine to the arc. The electrode cable forms one side of the circuit and runs from the electrode holder to the electrode terminal of the welding machine. The welding cable size is selected based on the maximum welding current used. Sizes range from AWG No. 6 to AWG No. 4/0, with amperage ratings from 75 amps upward. The work lead is the other side of the circuit and runs from the work clamp to the work terminal of the welding machine.

Covered electrodes, which become the deposited weld metal, are available in sizes from $\frac{1}{16}$ to $\frac{5}{16}$ in. in diameter and from 9 to 18 in. in length, with the 14-in. length the most popular. The covering dictates the usability of the electrode and provides the following:

- Gas shielding.
- Deoxidizers for purifying the deposited weld metal.
- Slag formers to protect weld metal from oxidation.
- Ionizing elements for smooth operation.
- Alloying elements to strengthen deposited weld metal.
- Iron powder to improve the productivity of the electrode.

The usability characteristics of different types of electrodes are standardized and defined by the American Welding Society (AWS). The AWS identification system indicates the characteristics and usability by classification numbers printed on the electrodes. Color code markings used for this purpose in the past are no longer employed.

Gas Metal-Arc Welding

Gas metal-arc (MIG) welding is an arc welding process that produces coalescence between a continous (consumable) filler-metal electrode and the work, by heating with an arc. The major components required for gas metal-arc welding are shown in Fig. 10-5. The electrode is in the form of a wire that is continuously and automatically fed into the arc to maintain a steady arc. This electrode wire, melted into the heat of the arc, is transferred across the arc and becomes the deposited welded metal. Shielding is obtained entirely from an externally supplied gas mixture. Fig. 10-6 shows the electrode wire, the gas shielding envelope, the arc, and the deposition of the weld metal.

Gas metal-arc welding yields top-quality welds in almost all metals and alloys. It also requires very little after-weld cleaning, can be completed at a relatively high speed, and produces little slag.

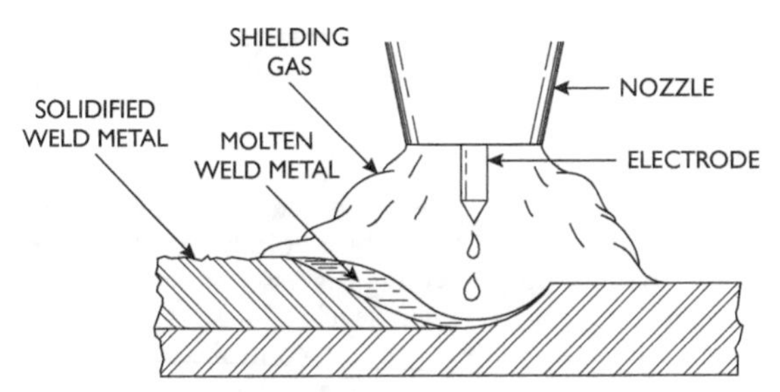

Fig. 10-5 Components for gas metal-arc welding.

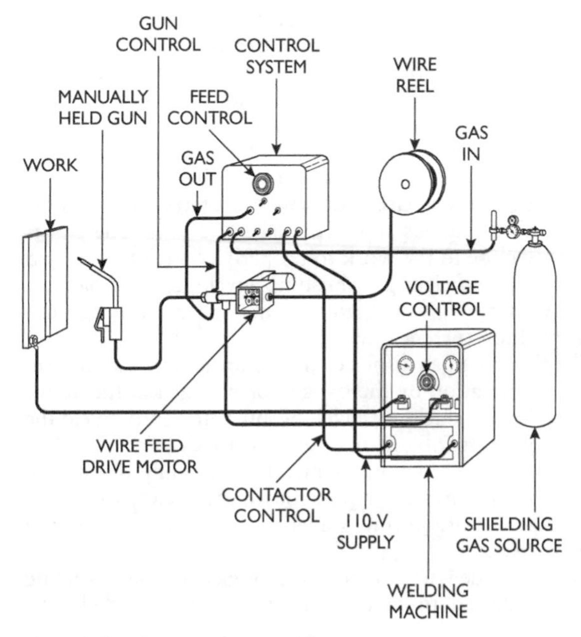

Fig. 10-6 Gas metal-arc welding.

Some of the process variations involve microwire for thin-gauge materials; CO_2 for low-cost, high-speed welding; and argon/oxygen for stainless steel.

Operation

The welding machine or power source for consumable-electrode welding is called a constant-voltage (CV) welder. This means that its output voltage is essentially the same with different welding current levels. These CV power sources don't have a welding current control and can't be used for welding with electrodes. The welding current output is determined by the load on the machine, which is dependent on the electrode wire-feed speed. The wire-feeder system must be matched to the constant-voltage power supply. At a given wire-feed speed rate, the welding machine will supply the proper amount of current to maintain a steady arc. Thus the electrode wire-feed rate determines the amount of welding current supplied to the arc.

The welding gun and cable assembly is used to carry the electrode wire, the welding current, and the shielding gas to the welding arc. The electrode wire is centered in the nozzle with the shielding gas supplied concentric to it. The gun is held fairly close to the work to properly control the arc and to provide an efficient gas shielding envelope. Welding guns for heavy-duty work at high currents and guns using inert gas at medium currents must be water cooled.

The shielding gas displaces the air around the arc to prevent contamination by the oxygen or nitrogen in the atmosphere. This gas shielding envelope must efficiently shield the area in order to obtain high-quality welds. The shielding gases normally used for gas metal-arc welding are argon, helium, or mixtures for nonferrous metals; CO_2 for steels; and CO_2 with argon and sometimes helium for steel and stainless steel.

The electrode wire composition for gas metal-arc welding must be selected to match the metal being welded. The electrode wire size depends on the variation of the process and the welding position. All electrode wires are solid and bare except in the case of carbon steel wire, which uses a very thin protective coating (usually copper).

Gas Tungsten-Arc Welding

Gas tungsten-arc (TIG) welding is a process that produces coalescence between a single tungsten (nonconsumable) electrode and the work, by heating with an arc. TIG was invented by the aircraft industry and is used extensively for hard-to-weld metals, primarily magnesium, aluminum, and also stainless steels. Shielding is obtained from an inert gas mixture, and filler metal may or may not be used. Fig. 10-7 shows the arc, the tungsten electrode, and the gas shielding envelope all properly positioned above the workpiece. The filler-metal rod is being fed manually into the arc and weld pool.

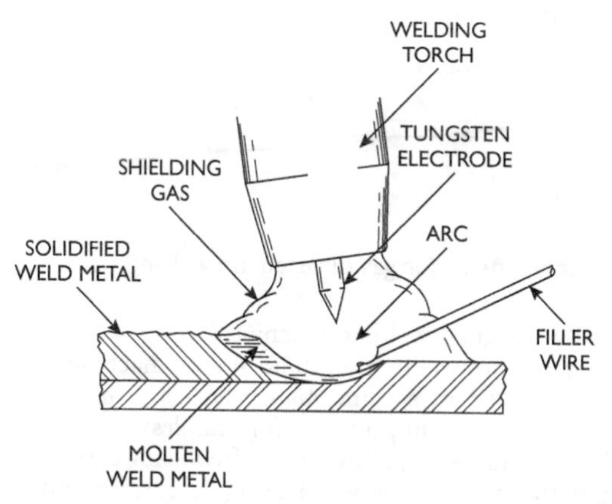

Fig. 10-7 Gas tungsten-arc welding.

Some of the outstanding features of gas tungsten-arc welding are top-quality welds in hard-to-weld materials and alloys; practically no after-weld cleanup; no weld spatter; and no slag production. The process can be used for welding aluminum, magnesium, stainless steel, cast iron, and mild steels. It will weld a wide range of metal thicknesses. The major components required for gas tungsten-arc welding are shown in Fig. 10-8.

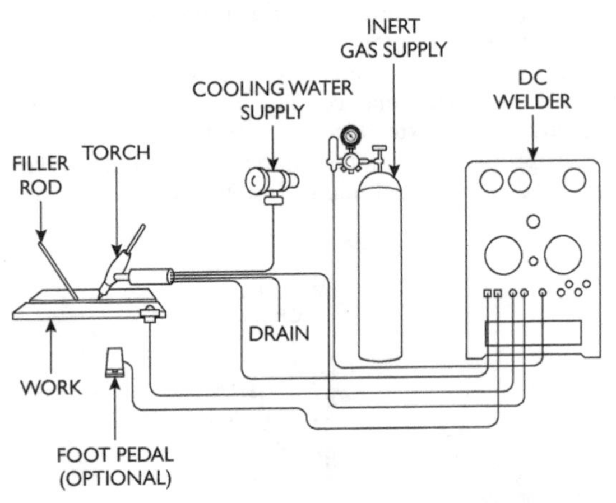

Fig. 10-8 Components for gas tungsten-arc welding.

A specially designed welding machine (power source) is used for tungsten-arc welding. Both ac and dc machines are built for the welding of specific materials—ac usually for aluminum and magnesium, and dc for stainless steel, cast iron, mild steel, and several alloys. High-frequency current is used to start the welding arc when using dc current and is used continuously with ac current. A typical gas tungsten-arc welding machine operates in a range of 3 to 350 amps and 10 to 35 V at a 60 percent duty cycle.

The torch holds the tungsten electrode and directs shielding gas and welding power to the arc. Most torches are water cooled, but some air-cooled torches are in use. The electrodes are made of tungsten and tungsten alloys. They have a very high melting point (6170°F) and are practically nonconsumable. The electrode does not touch the molten weld puddle; properly positioned, it hangs over the work and the arc keeps the puddle liquid. Electrode tips contaminated by contact with the weld puddle must be cleaned or they will cause a sputtering arc.

Filler metals are normally used except when very thin metal is welded. The composition of the filler metal should match that of the base metal. The size of the filler-metal rod depends on the thickness of the base metal and the welding current. Filler metal is usually added to the puddle manually, but automatic feed may be used on occasion. An inert gas—either argon, helium, or a mixture of both—shields the arc from the atmosphere. Argon is more commonly used because it is easily obtainable and, being heavier than helium, provides better shielding at lower flow rates.

Brazing

Brazing is a method of electrically joining copper conductors to other metals that are recognized by the National Electrical Code. It is seldom used, and then only for heavy grounding or similar connections. Because of the difficulty of this method, brazing has been largely replaced by exothermic welding.

Brazing is generally done with oxyacetylene welding equipment, but the metals to be joined don't melt together. Both metals are heated, and another metal is melted onto the two pieces of metal to join them.

Exothermic Welding

The exothermic welding process is most commonly known as a "Cadweld." By this process, the two (or in some cases, three or four) conductors to be joined are placed in a graphite mold. Then the welding material is added to the open area of the mold, called the *crucible.* The lid of the mold is then closed, and the welding material is ignited with a special tool. Instantly the mixture in the crucible melts and encloses the conductors in molten copper.

The process is fast, easy, and effective. There is very little danger of injury from the process as long as the operator follows the manufacturer's recommended method.

The only real drawback to the exothermic welding process is that a separate mold is required for each configuration of weld.

Safety

Welding involves several potentially hazardous conditions: very high temperature; use of explosive gases; and possible exposure to harmful light, toxic fumes, molten metal spatter, flying particles, and so forth. Welding hazards, however, can be successfully controlled to ensure the safety of the welder.

Some of the basic actions required are as follows:

- *Protective clothing* must be worn by the welder to shield skin from exposure to the brilliant light and heat given off by the arc (or flame).
- *A helmet* is required to protect the face and eyes from the arc.
- *Fire-resistant protective* clothing, shoes, leather gloves, jacket, apron, etc., are a necessity.
- *Dark-colored filter glass* in the helmet allows the welder to watch the arc while protecting the eyes.

Ventilation must be provided when welding in confined areas. The work area must be kept clean, and the equipment must be properly maintained.

Fire Watches

One of the more important safety requirements for field welding (welding on the jobsite, as opposed to welding in the shop) is that a fire watch must be maintained after the welding is completed.

The typical length of time required for the fire watch is the duration of the welding process and half an hour afterward. Thus, two people are required for every such welding job—one to actually perform the weld and a second to stand by with a fire extinguisher in hand, watching for the ignition of materials in the area.

It is important that you always allow enough time for fire watches in estimating the man-hours required for any installation. If you don't, your installations will nearly always take far longer than planned.

11. TRANSFORMERS

Transformers are devices that transform electrical energy from one circuit to another, usually at different levels of current and voltage but at the same frequency. This is done through electromagnetic induction in which the circuits never physically touch. The transformer is made of one or more coils of wire wrapped around a laminated iron core. Transformers come in many different sizes and styles.

Transformers are covered in *Article 450* of the National Electrical Code (NEC). Remember that as far as this article is concerned, the term transformer refers to a single transformer, whether it be a single-phase or a polyphase unit.

The requirements for installing and connecting transformers are detailed in *Article 450* of the NEC. They are as follows in this chapter.

Overcurrent Protection

Transformers operating at *over* 600 volts (V) must have protective devices for both the primary and secondary of the transformer, sized according to *Table 450.3(A)* of the NEC. If the specified fuse or circuit breaker rating does not correspond to a standard rating, the next larger size can be used. Transformers operating at over 600 V that are overseen only by qualified people are subject to somewhat different requirements, as is also shown in *Table 450.3(A)* of the NEC. If the specified fuse or circuit breaker rating does not correspond to a standard rating, the next larger size can be used.

Transformers rated 600 V *or less* can be protected by an overcurrent protective device on the primary side only; this must be rated at least 125 percent of the transformer's rated primary current. If the specified fuse or circuit breaker rating for transformers with a rated primary current of 9 amperes (A) or more does not correspond to a standard rating, the next larger size can be used. For transformers with rated primary currents of less than 9 amperes (A), the overcurrent device can be rated up to 167 percent of the primary rated current. (See Fig. 11-1.)

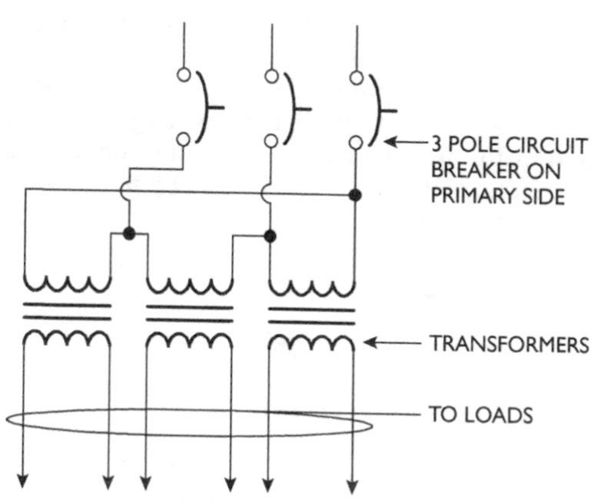

Fig. 11-1 Primary side overcurrent protection.

Transformers operating at *less than* 600 V are allowed to have overcurrent protection in the secondary only; this must be sized at 125 percent of the rated secondary current if the feeder overcurrent device is rated at no more than 250 percent of the transformer's rated primary current.

Transformers with thermal overload devices in the primary side don't require additional protection in the primary side unless the feeder overcurrent device is more than six times the primary's rated current (for transformers with 6 percent impedance or less) or four times primary current (for transformers with between 6 and 10 percent impedance). If the specified fuse or circuit breaker rating for transformers with a rated primary current of 9 A or more does not correspond to a standard rating, the next larger size can be used. For transformers with rated primary currents of less than 9 A, the overcurrent device can be rated up to 167 percent of the primary rated current.

Potential transformers must have primary fuses.

Autotransformers rated 600 V or less must be protected by an overcurrent protective device in each ungrounded input conductor; this must be rated at least 125 percent of the rated input current. If the specified fuse or circuit breaker rating for transformers with a rated input current of 9 A or more does not correspond to a standard rating, the next larger size can be used. For transformers with rated input currents of less than 9 A, the overcurrent device can be rated up to 167 percent of the rated input current. (See Fig. 11-2 and 11-3.)

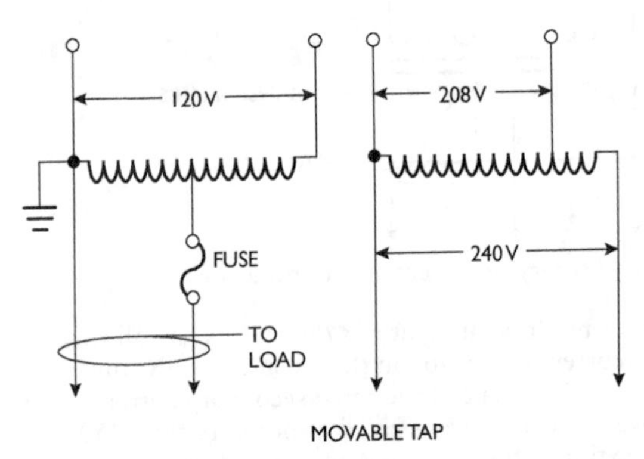

Fig. 11-2 Autotransformers.

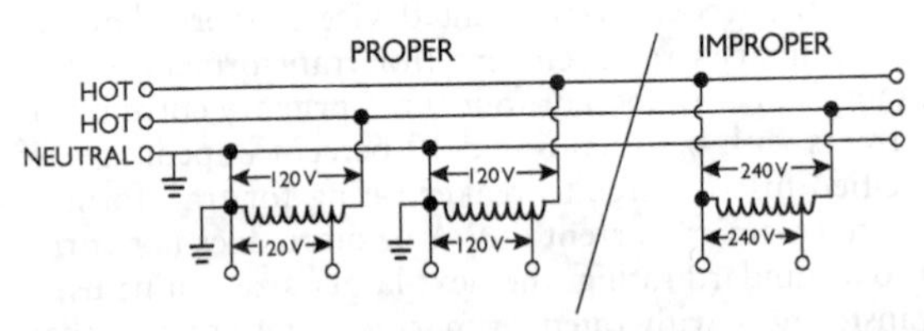

Fig. 11-3 Proper and improper autotransformer connections.

Installation

Transformers must be installed in places that have enough ventilation to avoid excessive heat build-up.

All *exposed noncurrent-carrying parts* of transformers must be grounded.

Transformers must be located in *accessible locations,* except as follows:

1. Dry-type transformers operating at less than 600 V that are located in the open on walls, columns, and structures don't have to be in accessible locations.
2. Dry-type transformers operating at less than 600 V and less than 50 volt-amperes are allowed in fire-resistant hollow spaces of buildings as long as they have enough ventilation to avoid excessive heating.

Indoor dry-type transformers rated 112½ kVA or less must be separated by at least 12 inches from combustible materials. Fire-resistant, heat-resistant barriers can be substituted for this requirement. Also, such transformers operating at 600 V or less that are completely enclosed are exempt from this requirement.

Indoor dry-type transformers rated over 112½ kVA must be installed in rooms made of fire-resistant materials. Such transformers with 80°C or higher ratings can be separated from combustible materials by fire-resistant, heat-resistant barriers or may be separated from combustible materials by at least 6 ft horizontally or 12 ft vertically. Also, transformers with 80°C or higher ratings that are completely enclosed (except for ventilated openings) are exempt from this requirement.

Materials cannot be stored in transformer vaults.

Transformer Connections

There are many, many different types of transformer connections. The most common connections used in electrical construction (as opposed to utility work) are shown here.

Single-Phase to Supply 120 V Lighting Load

As shown in Fig. 11-4, the transformer is connected between high voltage line and load with the 120/240 V winding connected in parallel. This connection is used where the load is comparatively small and the length of the secondary circuit is short. It is often used for a single customer.

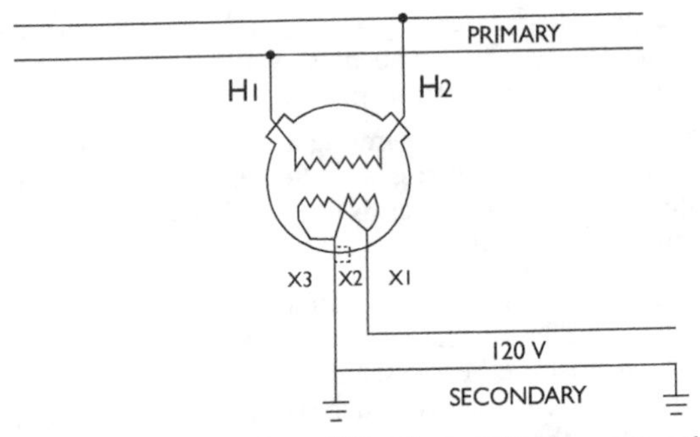

Fig. 11-4 Single-phase 120-volt transformer connection.

As shown in Fig. 11-5, the 120/240 V winding is connected in series and the midpoint is brought out, making it possible to serve both 120 V and 240 V loads simultaneously. This connection is used in most urban distribution circuits.

Single-Phase for Power

As shown in Fig. 11-6, the 120/240 V winding is connected in series serving 240 V on a 2-wire system. This connection is used for small industrial applications.

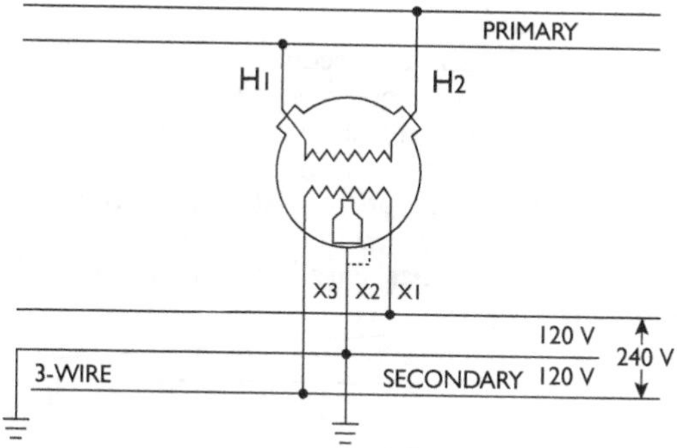

PRIMARY
H1
H2
X3
X2
X1
120 V
240 V
3-WIRE
SECONDARY
120 V

Fig. 11-5 Single-phase 120/240 V transformer connection.

PRIMARY
H1
H2
X3
X2
X1
SECONDARY

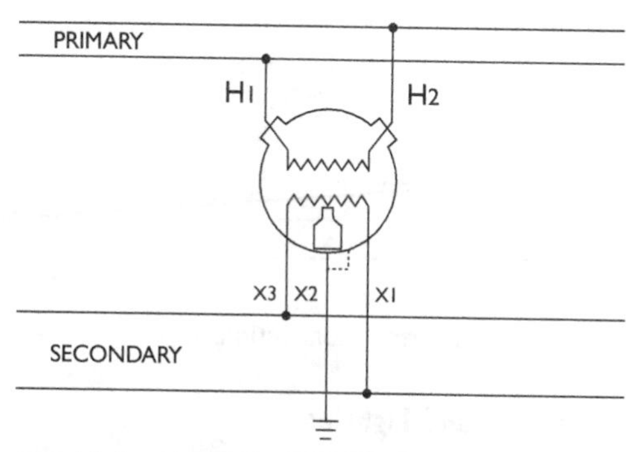

Fig. 11-6 Single-phase power connection.

Two-Phase Connections

As shown in Fig. 11-7, this connection consists merely of two single-phase transformers operated 90° out of phase. A 3-wire secondary is shown in this drawing, although in some cases a 4-wire or a 5-wire secondary is used.

Two-phase power was one of the first types of commercially generated electrical power, initially used at the Niagara Falls hydroelectric plant. It has since been replaced with 3-phase power in most places.

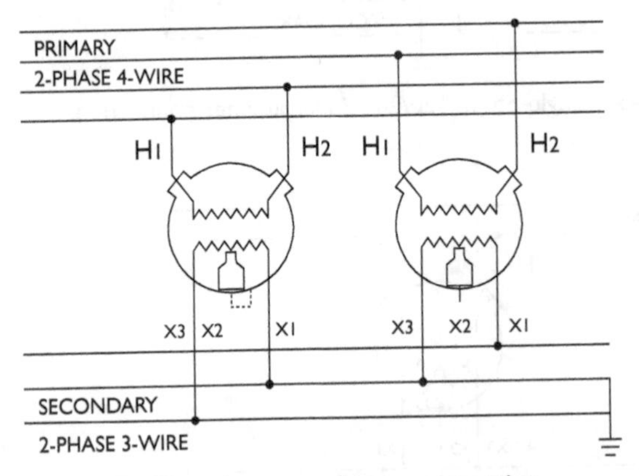

Fig. 11-7 Two-phase transformer connection.

Delta-Delta for Power and Lighting

As shown in Fig. 11-8, this connection is often used to supply a small single-phase lighting load and 3-phase power load simultaneously. As shown in the diagram, the mid-tap of the secondary of one transformer is grounded. Thus, the small lighting load is connected across the transformer with the mid-tap and the ground wire common to both 120 V circuits. The single-phase lighting load reduces the available 3-phase capacity. This type of connection requires special watt-hour metering.

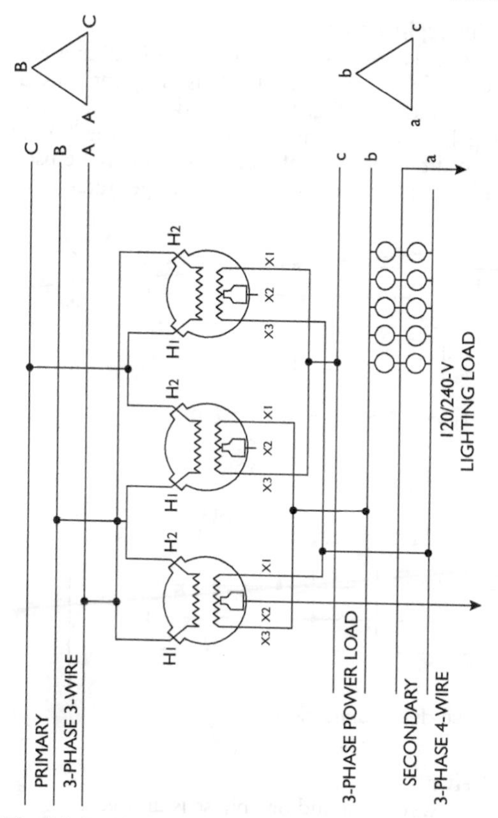

Fig. 11-8 Delta-delta transformer connection.

Open-Delta for Lighting and Power

Where the secondary load is a combination of lighting and power, the open-delta connected bank is frequently used. This connection (shown in Fig. 11-9) is used when the single-phase lighting load is large as compared with the power load. Here two different sizes of transformers may be used with the lighting load connected across the larger-rated unit.

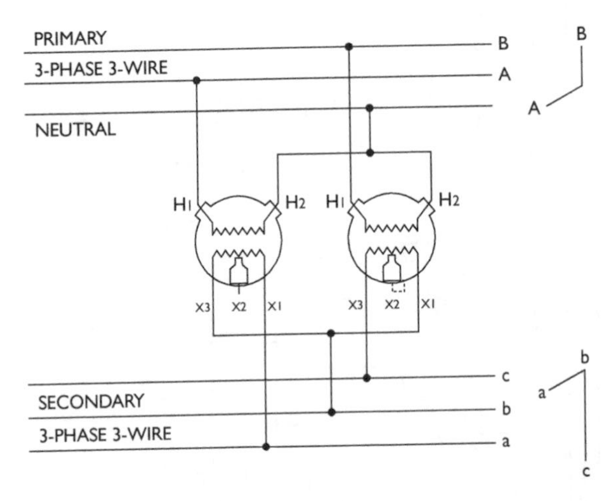

Fig. 11-9 Open-delta connection.

Open Wye-Delta

When operating wye-delta and one phase is disabled, service may be maintained at reduced load, as shown in Fig. 11-10. The neutral in this case must be connected to the neutral of the setup bank through a copper conductor. The system is unbalanced, electrostatically and electromagnetically, so that telephone interference may be expected if the neutral is connected to ground. The useful capacity of the open-delta, open wye bank is 87 percent of the capacity of the installed transformers when the two units are identical.

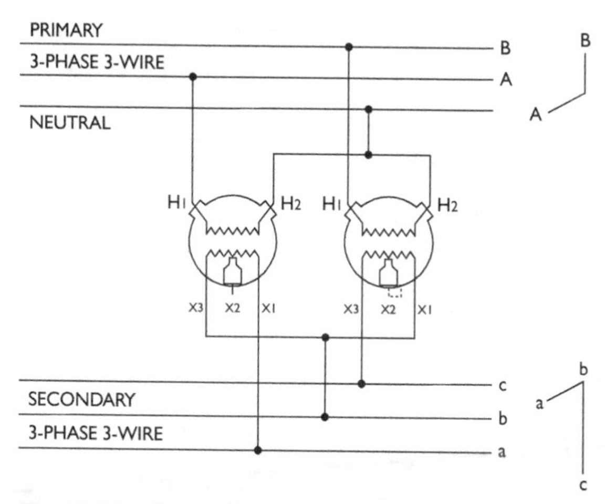

Fig. 11-10 Open wye-delta connection.

Delta-Wye for Lighting and Power

To eliminate unbalanced primary currents, the delta-wye connection (shown in Fig. 11-11) is often used. Here the neutral of the secondary 3-phase system is grounded, and the single-phase loads are connected between the different phases of wires and the neutral while the three-phase loads are connected to the phase wires. Thus, the single-phase load can be balanced on three phases in each transformer bank; and banks may be paralleled if desired.

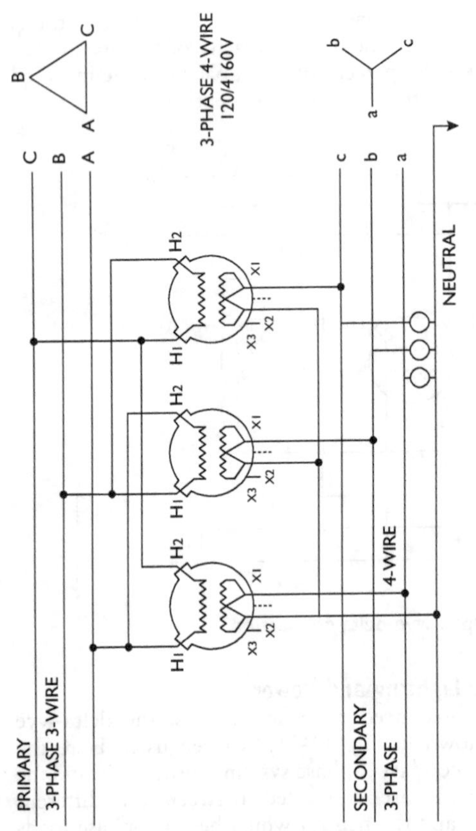

Fig. 11-11 Delta-wye connection.

Wye-Wye for Lighting and Power

Fig. 11-12 shows a system on which the primary voltage was increased from 2400 to 4160 V to increase the potential capacity of the system. The previously delta-connected distribution transformers are now connected from line to neutral. The secondaries are connected in wye. In this system, the primary neutral is connected to the neutral of the supply voltage through a metallic conductor and is carried with the phase conductor to minimize telephone interference. If the neutral of the transformer is isolated from the system neutral, an unstable condition caused primarily by third harmonic voltages results at the transformer neutral. If the transformer neutral is connected to ground, the possibility of telephone interference is greatly enhanced, and there is also a possibility of resonance between the line capacitance to ground and the magnetizing impedance of the transformer.

Wye-Wye Autotransformers with 3-Phase, 4-Wire System

When the ratio of transformation from the primary to secondary voltage is small, the most economical way of stepping down voltage is by using autotransformers as shown in Fig. 11-13. For this application, it is necessary that the neutral of the autotransformer bank be connected to the system neutral.

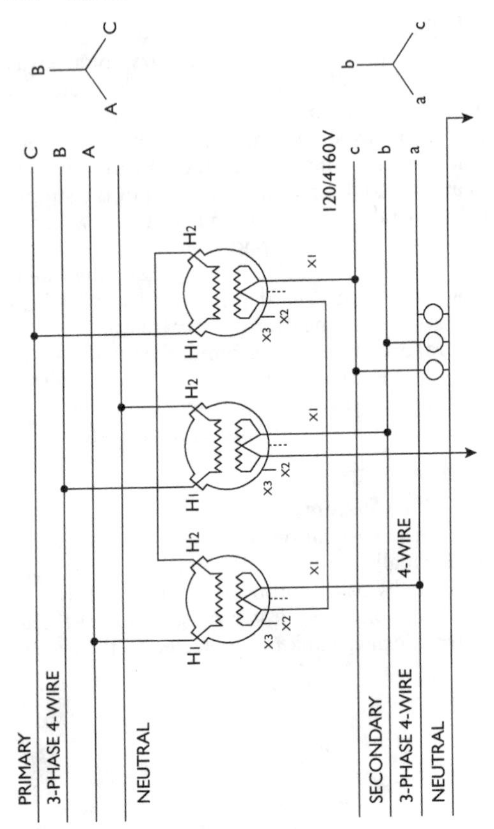

Fig. 11-12 Wye-wye connection.

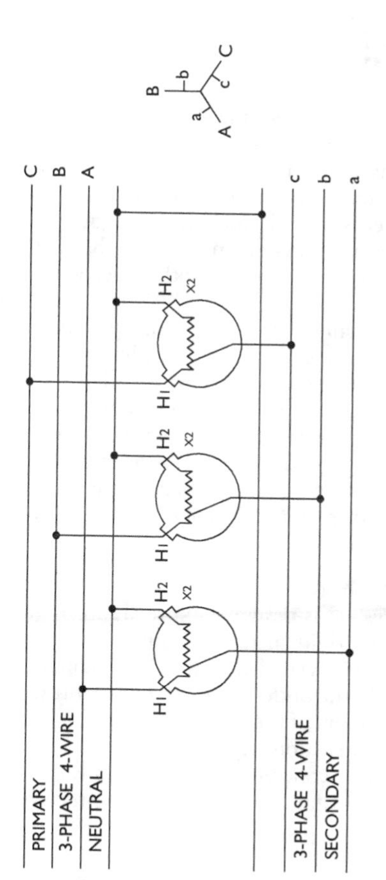

Fig. 11-13 Autotransformer connection.

12. CIRCUIT WIRING

By far the most common type of wiring necessary for most electrical installations is branch circuit wiring. Services and feeders are the big items, but far more time and effort are spent on branch circuits than on anything else.

The two main concerns with branch circuitry are first that it be installed according to code and second that it be installed in a reasonable amount of time. Both topics will be addressed in this chapter.

Branch circuit wiring will be defined as that portion of the electrical system covering such items as outlet boxes of all types, all 1-inch and smaller conduits with fittings, and types of cable and building wire up to and including No. 8 AWG. In general, branch circuits include all wiring for lighting, receptacles, small power, and communication systems.

Specific items to be covered include branch circuit rough-in, surface metal raceway systems, branch circuit wiring, busway, and branch circuit cable.

The Branch Circuit Rough-in

An electrical raceway system is designed expressly for holding wires. In addition to rigid metallic conduit, electrical metallic tubing (EMT), or polyvinyl chloride (PVC) conduit, a raceway system includes the outlet boxes and other fittings through which the conductors of the system will be installed. For general building construction, rigid or PVC conduit is normally used in and under concrete slabs, whereas EMT is used for all above-surface installations except where the system will be exposed to severe mechanical injury.

Rigid Nonmetallic Conduit

Rigid nonmetallic conduit and fittings (PVC electrical conduit) may be used where the potential is 600 V or less in direct earth burial not less than 18 in. below the surface. If less than 18 in., the PVC conduit must be encased in not less than 2 in. of concrete.

PVC conduit may further be used in walls, floors, ceilings, cinder fill, and in damp and dry locations except in certain hazardous locations for support of fixtures or other equipment and where subject to physical damage.

PVC conduit can be cut easily at the job site without special tools. Sizes ½ through 1½ in. can be cut with a fine-tooth hand saw. For sizes 2 through 6 in., a miter box or similar saw guide should be used to keep the material steady and assure a square cut. In order to assure satisfactory joining, care should be taken not to distort the end of the conduit when cutting.

After cutting, deburr the ends of the pipe and wipe them clean of dust, dirt, and plastic shavings. Deburring is easily done with either a pocket knife or hand file. EMT is sometimes deburred with a pair of pliers (carefully used).

One of the more important advantages of PVC conduit, as compared to metal conduit, is the ease and speed at which the solvent-cemented joints can be made. The steps used for a good joint are as follows:

1. Wipe the conduit clean and dry.
2. Apply a full, even coat of PVC cement to the end of the conduit. (Most cans of solvent come with a built-in applicator.)
3. Push the conduit into the fitting with a slight twisting motion until it bottoms, and then rotate the conduit in the fitting to evenly distribute the cement. Too much cement will find its way to the inside surfaces of the conduit and will reduce the internal cross-sectional area.

These operations should not take more than 20 seconds. Joints generally take several minutes to dry completely.

Once the proper amount of cement has been applied, a bead of cement will form at the joint. Wipe the joint with a cloth or brush to remove the excess cement. (See Fig. 12-1.)

Size Fitting(in.)	Pint No. of Joints	Quart No. of Joints
½	350	700
¾	200	400
1	150	300
1¼	110	220
1½	80	160
2	45	90
3	35	70
3½	30	60
4	25	50
5		25
6		15

Fig. 12-1 Joining PVC conduit.

Although various types of premade bends are available, most bends in PVC conduit are made on the job. PVC bends are quite a bit easier to perform than bends in metal conduit, especially since almost no calculations are required for these bends. The conduit must be evenly heated along the entire length of the bend. The most common method of heating the PVC conduit is by use of a small hot box. These devices employ infrared heat that is quickly absorbed by the conduit. Small sizes of PVC are generally ready to be bent after less than a minute in the heater. Larger diameters require two or three minutes depending on conditions. Other methods of heating PVC conduit for bending include heating blankets and hot-air blowers. Immersion in hot liquids (about 275°F) is also satisfactory. The use of torches or other flame-type devices is not recommended. PVC conduit exposed to excessively high temperatures may take on a brownish color. Sections showing evidence of such scorching should be discarded.

A stub, as shown in Fig. 12-2, is normally formed at the end of a conduit section with a rise of between 12 and 18 in. to an outlet box. Stubs are among the easiest of bends.

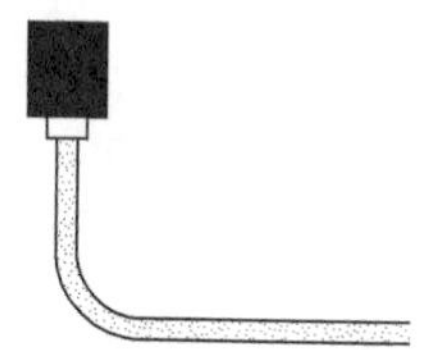

Fig. 12-2 Conduit stub.

A saddle, as shown in Fig. 12-3, is used to run the conduit over or around an obstruction, such as a beam or pipe.

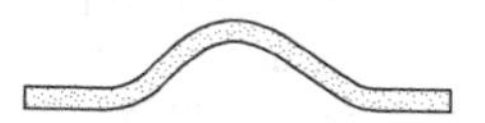

Fig. 12-3 Conduit saddle.

Offsets, as shown in Fig. 12-4, are changes in the line of conduit run in order to avoid an obstruction or to meet an opening in a box or enclosure.

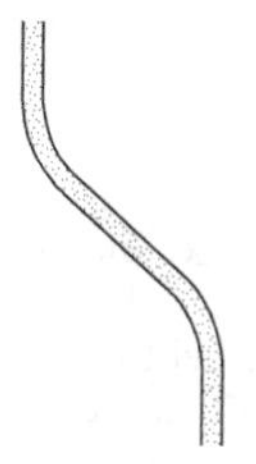

Fig. 12-4 Conduit offset.

Concentric bends, as shown in Fig. 12-5, are a series of conduits turning together around a common center point. Each requires a different bending radius. This process, which is very difficult with metal conduits, is relatively simple with PVC conduit.

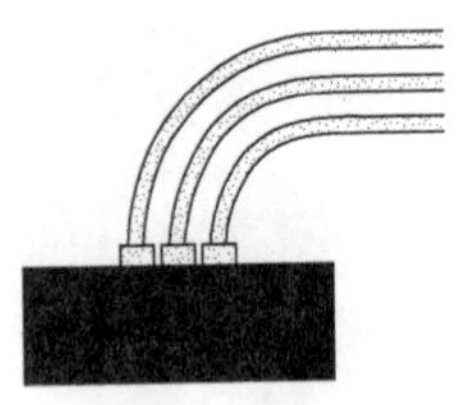

Fig. 12-5 Conduit concentric bends.

If a number of identical bends are required, a jig can be helpful. (See Fig. 12-6.) A simple jig can be made by sawing a sheet of plywood to match the desired bend. Nail this piece to a second sheet of plywood. The heated conduit section is placed in the jig, is sponged with water to cool, and is then ready to install. Care should be taken to fully maintain the inside diameter (ID) of the conduit when handling.

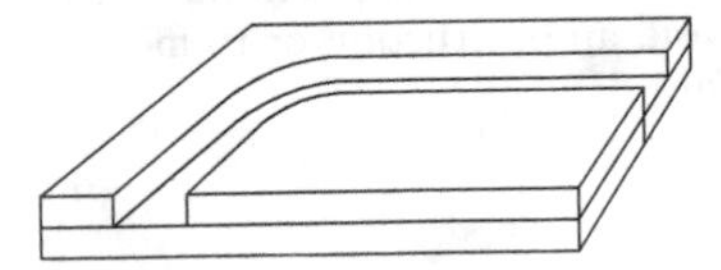

Fig. 12-6 Bending jig.

If only a few bends are needed, scribe a chalk line on the floor or workbench. Then match the heated conduit to the chalk line and let cool. The conduit must be held in the desired position until relatively cool, since the PVC material will tend to go back to its original shape.

Another method is to take the heated conduit section to the point of installation and form it to fit the actual installation with the hands. (See Fig. 12-7.) Then wipe a wet rag over the bend (Fig. 12-8) to cool it. This method is especially effective in making "blind" bends or compound bends.

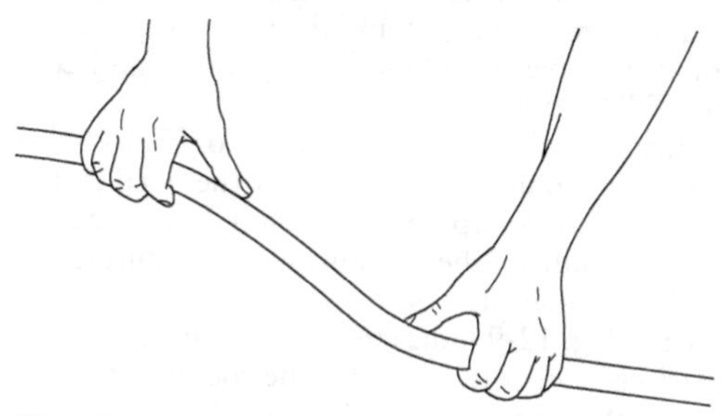

Fig. 12-7 Heat-bending PVC conduit.

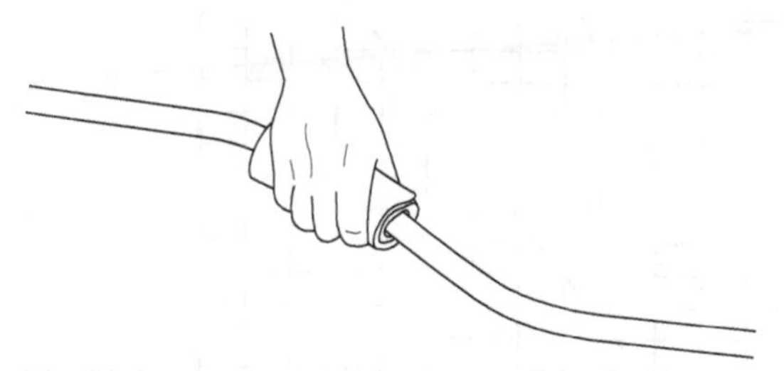

Fig. 12-8 Cooling heated PVC conduit.

Bends in small-diameter PVC conduit (½ to 1½ in.) require no filling for code-approved radii. In larger sizes, for other than minor bends, the interior must be supported to prevent collapse. Use a flexible spring or air pressure.

For the best possible bends (without crimping the conduit), place airtight plugs in each end of the conduit section before heating. The retained air will expand during the heating process and will hold the conduit open during the bending. Don't remove the plugs until the conduit is cooled.

In applications where the conduit installation is subject to constantly changing temperatures and the runs are long, precautions should be taken to allow for expansion and contraction of the PVC conduit.

When expansion and contraction are factors, an O-ring expansion coupling should be installed near the fixed end of the run, or fixture, to take up any expansion or contraction that may occur. Confirm the expansion and contraction length available in these fittings; it may vary by manufacturer. The chart in Fig. 12-9 indicates what expansion can be expected at various temperature levels. The coefficient of linear expansion of PVC conduit is 0.0034 in./10 ft/°F.

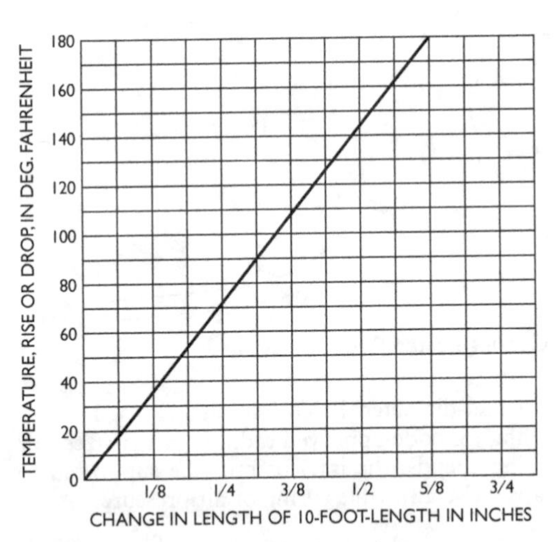

Fig. 12-9 Expansion of PVC conduit.

Expansion couplings are seldom required in underground or slab applications. Expansion and contraction may generally be controlled by bowing the conduit slightly or by immediate burial. After the conduit is buried, expansion and contraction cease to be factors. Care should be taken, however, in constructing a buried installation. If the conduit must be left exposed for an extended period of time during widely variable temperature conditions, allowance should be made for expansion and contraction.

In above-ground installations, care should be taken to provide proper support of PVC conduit due to its semirigidity. This is particularly important at high temperatures. Distance between supports should be based on temperatures encountered at the specific installation. The chart in Fig. 12-10 clearly outlines at what intervals support is required for the PVC conduit at various temperature levels.

Recommended Conduit Support Spacing							
Nominal	***Maximum Temperature in Degrees Fahrenheit***						
Size(in.)	***20° ft.***	***60° ft.***	***80° ft.***	***100° ft.***	***120° ft.***	***140° ft.***	***160° ft.***
½	4	4	4	4	2½	2½	2
¾	4	4	4	4	2½	2½	2
1	5	5	5	5	3	2½	2½
1¼	5	5	5	5	3	3	2½
1½	5	5	5	5	3½	3	2½
2	5	5	5	5	3½	3	2½
2½	6	6	6	6	4	3½	3
3	6	6	6	6	4	3½	3
3½	7	7	7	6	4	3½	3½
4	7	7	7	7	4½	4	3½
5	7	7	7	7	4½	4	3½
6	8	8	8	8	5	4½	4

Fig. 12-10 Recommended outdoor supports for PVC conduit.

Rigid Metal Conduit

Galvanized rigid metal conduit may be used under almost all atmospheric conditions and occupancies. However, in areas subject to severe corrosive influences, rigid metal conduit and fittings should not be installed unless corrosion protection is provided.

It has sometimes been the practice to use special conduit cutters to cut the conduit during the installation of rigid conduit. This should be avoided, since the cutters leave a large burr and hump inside the conduits. Then additional time is required to remove the burr. A better method is to use a lightweight portable electric hacksaw using blades with 18 teeth per inch.

Conduit cuts should be made square and the inside edge of the cut adequately reamed to remove any burr or sharp edge that might damage the insulation of the conductors when they are pulled in later. Lengths of conduit should be accurately measured before they are cut—recutting will obviously result in lost time.

A vise stand that will securely hold the conduit and not shift about as cuts are being made should be provided for each crew. A power hacksaw of either the blade or band type should be provided, as well as a sufficient supply of hacksaw blades to be used in the electricians' hacksaw frames. The cut should be made entirely through the conduit and not broken off the last fraction of an inch. Although the hacksaw may be used to cut smaller sizes of conduit by hand, the larger sizes should not be cut by hand except in emergencies. Not only will hand cutting of the large sizes take up too much time, but it is also extremely difficult to cut such large sizes of conduit square.

In addition to the length needed for the piece of conduit for a given run, an additional ⅜ in. should be allowed on smaller sizes of conduit for the wall of the box and bushing. Because larger sizes of conduit are usually connected to heavier boxes, allow approximately ½ in. If additional locknuts are needed in a run, a ⅛-in. allowance for each locknut will usually be sufficient. Where conduit bodies are used, include the length of the threaded hub in the measurements.

The usual practice for threading the smaller sizes of rigid conduit is to use a pipe vise in conjunction with a die stock with proper size guides and sharp cutting dies properly adjusted and securely held in the stock. Clean, sharp threads can be cut only when the conduit is well lubricated; use a good lubricant and plenty of it.

Conduits should always be cut with a full thread. To accomplish this, run the die up on the conduit until the conduit just about comes through the die for about one full thread. This gives a thread length that is adequate for most purposes. However, don't overdo it; if the thread is too long, the portion that does not fit into the coupling will corrode because threading removes the protective covering.

Clean, sharply cut threads also make a better continuous ground and save unnecessary labor. It takes little extra time to make certain that threads are properly made, but a little extra time spent at the beginning of a job may save much time later on.

On projects where a considerable number of relatively short sections of conduit will be required for nipples, considerable threading time can be saved by periodically gathering up the short lengths of conduit resulting from previous cuts and reaming and running a thread on one end with a power threader and redistributing these lengths to the installation points about the job. This procedure eliminates one-hand threading operations in many instances.

When threadless couplings and connectors are used with rigid conduit, they should be made tight. Where the conduit is to be buried in concrete, the couplings and connectors must be of the concrete type; where used in wet locations, they must be of the rain-tight type.

The installation of rigid conduit for branch circuit raceways often requires many changes of direction in the runs, ranging from simple offsets to complicated angular offsets, saddles, and so on.

In bending elbows, care should be taken to comply with the National Electrical Code (NEC). In general, the Code states that an elbow or 90° bend must have a minimum radius of six times the inside diameter of the conduit. Therefore, the radius of a 2-in. conduit must have a radius of at least 12 in., a 3-in. conduit 18 in., and so forth.

Bends in the smaller sizes of conduit are normally made by hand with the use of conduit hickeys or benders. In some cases, where many bends of the same type must be made, hand roller "Chicago benders" or hydraulic benders are used to simplify making the bends to certain dimensions.

Occasionally, rigid conduit will have to be rebent after it is installed. In such cases, the rebending must be done carefully so that the conduit does not break or crimp. Most often, these rebinds will have to be made at stubups—conduits emerging through concrete floors. To rebend conduit, the concrete should be chipped away for a few inches around the conduit, and then the conduit should be warmed with a propane torch. It can then be bent into the required shape without further trouble.

Electrical Metallic Tubing

Electrical metallic tubing may be used for both exposed and concealed work, except where the tubing will be subjected to severe physical damage, such as in cinder concrete, unless the tubing is at least 18 in. under the fill.

The tubing should be cut with a hand or power hacksaw using blades with 24 or 32 teeth per inch, after which the cut ends should be reamed to remove all rough edges. Threadless couplings and connectors used must be made tight, and the proper type should be used for the situation; that is, concrete-tight types should be used when the tubing is buried in concrete and rain-tight type used when installed in a wet location. Supports must be provided, when installed above grade, at least every 10 ft and within 3 ft of each outlet box or other termination point.

Bends are made in EMT much the same as for rigid conduit. However, roll-type benders are used exclusively for

EMT. This type of bender has high supporting side walls to prevent flattening or kinking of the tubing and a long arc that permits the making of 90° bends (or any lesser bends) in a single sweep.

Many time-saving tools and devices have appeared on the market during the past years to facilitate the installation of EMT. Table hydraulic speed benders, for example, make 90° bends or offsets in 5–10 seconds. Shoes are available for ½- through 1¼-in. EMT. Mechanical benders are also available for sizes through 2-in. EMT.

An automatic "kicker" will eliminate the need for offset connectors wherever ¾- or ½-in. EMT is used. The end of a piece of EMT is inserted in the blocks of the device, and one push of the handle (about a 2-second duration) makes a perfect offset. Every bend is identical; this eliminates lost time refitting or cutting and trying.

The EMT hand benders with built-in degree indicators let the operator make accurate bends between 15° and 90° more quickly, since in-between measurements are eliminated.

Surface Metal Raceway

Surface metal raceway is one of the exposed wiring systems that is quite extensively used in existing buildings where new wiring or extensions to the old are to be installed. Although it does not afford such rugged and safe protection to the conductors as rigid conduit or EMT, it is a very economical and quite dependable system when used under the conditions for which it was designed. The main advantage of surface metal raceway is its neat appearance where wiring must be run on room surfaces in finished areas.

Surface metal raceways may be installed in dry locations except where subject to severe physical damage or where the voltage is 300 V or more between conductors (unless the thickness of the metal is not less than 0.040 in). Furthermore, surface metal raceways should not be used in areas that are subjected to corrosive vapors, in hoistways, or in any hazardous location. In most cases, this system should not be used for concealed wiring.

Various types of fittings for couplings, corner turns, elbows, outlets, and so on, are provided to fit these raceways. Fig. 12-11 shows a number of the most common fittings in use.

Many of the rules for other wiring systems also apply to surface metal raceways. That is, the system must be continuous from outlet to outlet, it must be grounded, all wires of the circuit must be contained in one raceway, and so on.

In planning a surface metal raceway system, the electrician should make certain that all materials are present before the installation is begun. One missing fitting could hold up the entire project.

Proper tools should also be provided to make the installation easier. For example, a surface metal raceway bending tool will enable electricians to bend certain sizes of the raceway like rigid conduit or EMT. With such a tool, many of the time-consuming fittings can be eliminated because the tool can make 90° bends, saddles, offsets, back-to-back bends, and so on. Some bending tips for the Wiremold Benfield Bender follow:

1. Bending with the tool in the air—Apply hand pressure as close to the tool as possible. Keep pressure close to the groove for smoother bends and greater accuracy.
2. Bending on the floor—Work on hard surfaces. Avoid soft sand or deep pile carpets.
3. Degree scale—One side of the tool (closed hook side) is calibrated for 500 wiremold. The opposite, open hook side, is scaled for 700.
4. Zero-degree (0°) line—The zero-degree line in the bottom of the groove adjacent to the hook is the point of beginning of the bend.
5. Rim notches—The rim notches closest to the hook indicate the exact center of a 45° bend. Numerals "500" and "700" tell the operator which notch to use for which size wiremold.

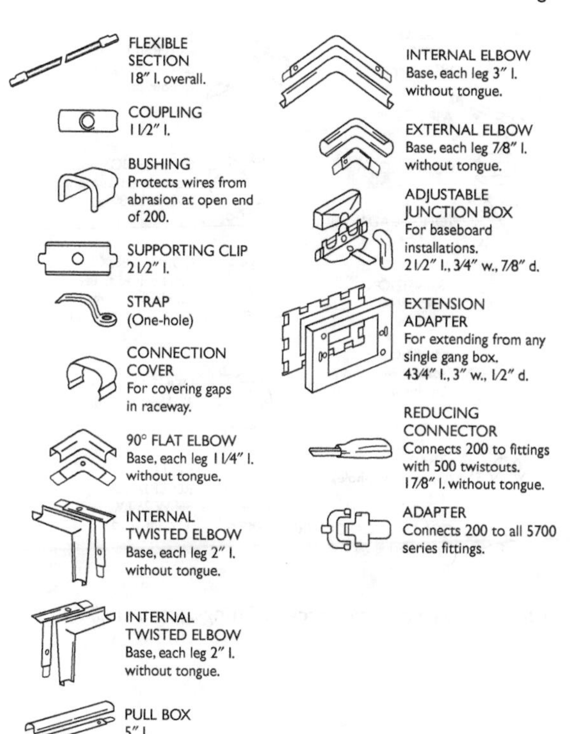

Fig. 12-11 Surface metal raceway fittings. *(continues)*

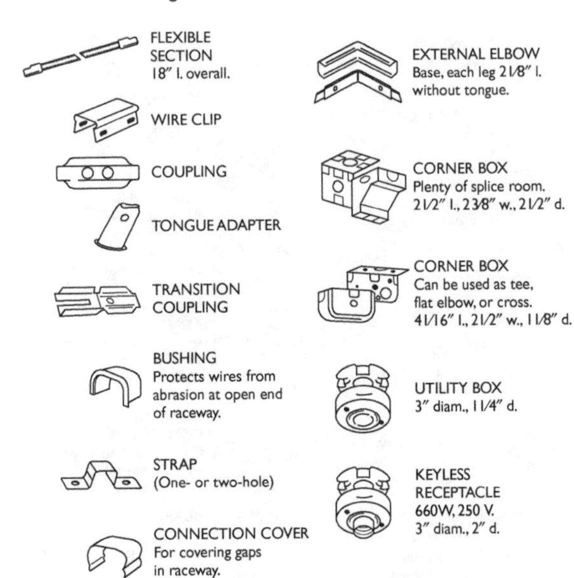

Fig. 12-11 Surface metal raceway fittings. *(continued)*

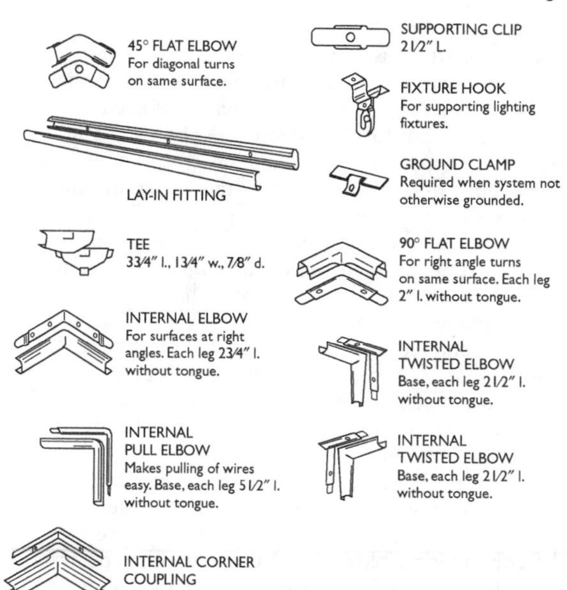

Fig. 12-11 Surface metal raceway fittings. *(continued)*

Branch Circuit Wiring

In most cases, the installation of branch circuit wires is merely routine. However, there are certain practices that can reduce labor and materials to the extent that such practices should be given careful consideration. The use of modern equipment, such as vacuum fish tape systems, is another way to reduce labor during this phase of the wiring installation. The proper size and length of fish tape, as well as the type, should be one of the first considerations. For example, if most of the runs between outlets are only 20 ft or less, a short fish tape of, say 25 ft, will easily handle the job and won't have the weight and bulk of a larger one. When longer runs are encountered, the required length of fish tape should be enclosed in one of the metal or plastic fish tape reels. This way the fish tape can be rewound on the reel as the pull is being made so as to avoid having an excessive length of tape lying around on the floor.

When several bends are present in the raceway system, the insertion of the fish tape may be made easier by using flexible fish tape leaders on the end of the fish tape.

The combination blower and vacuum fish tape systems are ideal for use on long runs and can save much time. Basically, the system consists of a tank and air pump with accessories. An electrician can vacuum (Fig. 12-12) or blow (Fig. 12-13) a line or tape in any size conduit from ½ through 4 in. or even 5- and 6-in. conduits with optional accessories.

After the fish tape is inserted in the raceway system, the wires must be firmly attached by some approved means. On short runs, where only a few conductors are involved, all that is necessary is to strip the insulation from the ends of the wires, bend these ends around the hook in the fish tape, and securely tape them in place. Where several wires are to be pulled together over long and difficult conduit runs (with several bends), the wires should be staggered and the fish tape securely taped at the point of attachment so that overall diameter is not increased any more than is absolutely necessary. Staggering is done by attaching one wire to the fish tape and then attaching the second wire a short distance behind this to the bare copper conductor of the first wire. The third wire, in turn, is attached to the second wire and so forth.

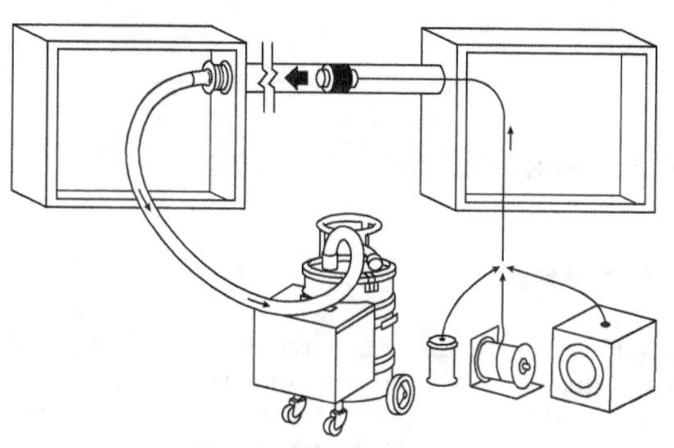

Fig. 12-12 Conduit vacuum system.

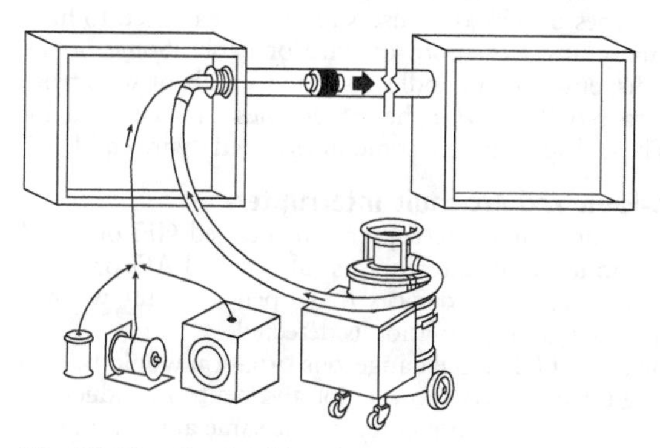

Fig. 12-13 Conduit blow line system.

Basket grips (Fig. 12-14) are available in many sizes for almost any size and combination of conductors. They are designed to hold the conductors firmly to the fish tape and can save the electrical workers much time and trouble that would be required when taping wires.

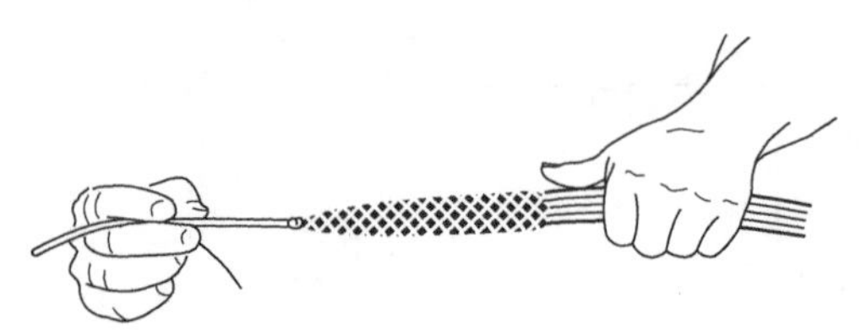

Fig. 12-14 Basket grip.

In all but very short runs, the wires should be lubricated with a good quality wire lubricant prior to attempting the pull and also during the pull. Some of this lubricant should also be applied to the inside of the conduit itself.

Wire dispensers are great aids in keeping the conductors straight and in facilitating the pulling of conductors. Many different types of wire dispensers are now marketed to handle virtually any size spool of wire or cable. Some of the smaller dispensers can handle up to ten spools of wire from No. 22 to No. 10 AWG; the larger ones can handle a lot more. These dispensers are sometimes called "wire caddies."

Ground-Fault and Arc-Fault Interrupters

Both ground-fault interrupters (also called GFI or GFCI devices) and arc-fault interrupters (also called AFI or AFCI devices) are automatic devices that open a circuit when a potentially dangerous situation is detected.

In the case of GFIs, the dangerous situation would be current being unequal between the hot and neutral conductors. If these conductors are not carrying the same amount of current, it is evident that some current is taking some path other than the appropriate one. In such a case, the GFI interrupts the circuit, and injuries are avoided.

GFIs are required to protect all 125-V, 15- and 20-amp receptacles in dwellings. They must be used in the following areas in dwellings (use in other areas is optional): bathrooms, garages, outdoor crawl spaces, unfinished basements, kitchens (counter-top receptacles), wet bar sinks, and boathouses. In locations other than dwellings, all 125-V, 15- and 20-amp receptacles in bathrooms, kitchens, and rooftops must be protected. See *Section 210.8* of the NEC for details.

AFIs sense electronic characteristics that are present in circuits that are experiencing electrical arcs. Arcs are common sources of ignition—that is, they start fires. So, the purpose of the AFCI is to sense an arc, and to interrupt the circuit before a fire can begin.

AFIs are required to protect all 15- or 20-amp, 125-volt circuits (that is, standard circuits) in bedrooms in homes. See *Section 210.12* of the NEC for details.

Note that the Code requirements are for GFIs to protect *receptacles*, and for AFIs to protect *circuits*.

Branch Circuit Cables

Nonmetallic-sheathed (type NM) and metal-clad (type AC) cables are very popular for use in residential and small commercial wiring systems. In general, both types of cable may be used for both exposed and concealed work in normally dry locations. They may be installed or fished in air voids in masonry block or tile walls, where such walls are not exposed or subject to excessive moisture or dampness. Type NM cable can't be installed where it can be exposed to corrosive fumes or vapors; nor can it be embedded in masonry, concrete, fill, or plaster; nor can it run in shallow chase in masonry or concrete and covered with plaster or a similar finish.

Type NM cable may not be used as a service-entrance cable, in commercial garages, in theaters and assembly halls, in motion picture studios, in storage battery rooms, in hoistways, or in any hazardous location; it may not be embedded in poured cement, concrete, or aggregate. Type AC cable has the same restrictions.

For use in wood structures, holes are bored through wood studs and joists first, and then the cable is pulled through these holes to the various outlets. Normally, the holes give sufficient support, providing they are not over 4 ft on center. Where no stud or joint support is available, staples or some similar supports are required for the cable. The supports must not exceed 4¼ ft; cable must also be supported within 12 in. of each outlet box or other termination point.

Proper tools facilitate the running of branch circuit cables and include such items as sheathing strippers for stripping the NM cable; hacksaw for cutting and stripping type AC (BX) cable; and a carpenter's apron for holding staples, crimp connectors, and wire nuts.

Conduit Installation

There are a number of techniques that can be used to speed up the installation of conduit.

Any percentage of your conduit installation labor that you can save is very important, because the raceway installation (which is mostly conduit) accounts for the majority of the labor on your jobs.

One of the first steps toward speeding conduit installations is to have your ladders set up as work stations, as shown in Fig. 12-15. The left-hand sketch shows a typical 6-ft step ladder set up as a work station. There are two small notches cut in the side rails of the ladder, just above the second step. (Where OSHA restrictions prohibit this method, special add-on brackets can be used instead.) On the right side of the ladder, there are two hooks. One hook should be on the front side rail for a hacksaw. The other hook should be mounted on the rear side rails (the side that usually has rungs rather than steps) for hanging a conduit bender. Another hook should be mounted on the left side of the ladder for a canvas bag. In this bag will go various small parts that are needed for the installation; typically these would be fittings, straps, or fasteners. (For left-handed workers, this set-up should be just the opposite.) By doing this, all of the necessary items for the installation are arranged to be at the electrician's fingertips, rather than scattered around the job site. If fairly large hooks are used, the electrician can pick up the ladder, with all its appurtenances hanging from it, and move from one location to another quickly and easily.

The sketch on the right in Fig. 12-15 shows how to use the notches in the side rails for cutting conduit. These notches should be cut in a triangular shape. One side should be flush with the second step of the ladder. The conduit to be cut should be placed into these notches. The electrician can then place his or her left knee on the conduit (assuming that the electrician is right-handed) and cut the conduit easily.

Fig. 12-15 Ladders used as workstations.

Slabbing

Slabbing simply means running conduits underneath the floor slab of a building. Running conduits in the floor slab rather than overhead saves quite a bit of money on both material and labor. The reasons are as follows:

1. Running the conduit in the slab uses up to 50 percent less conduit than it takes to run the same circuits overhead. Running from one receptacle to another requires only two 2-ft vertical runs of conduit between outlets, for a total vertical length of 4 ft in each receptacle-to-receptacle run. By going overhead, a minimum vertical drop of 7 ft is required to each outlet, making the total vertical length for the same receptacle-to-receptacle run 14 ft, rather that 4 ft. Additionally, by running under slab, you can usually get away with a more direct (angled, rather than straight and perpendicular) routing of the conduits.
2. Conduits run under the slab need almost no strapping. This eliminates a great deal of the expense for both material and labor for straps and fasteners. (Fasteners are screws, bolts, nails, anchors, and so on.)
3. PVC conduit is usually used for running under the slab, rather than EMT. This provides a savings in material cost, because PVC is cheaper than EMT. Additionally, PVC comes with integral couplings, saving both material and labor expense.

There are two drawbacks with slabbing: One is that it requires you to install a separate ground wire, which you may not have had to do using metal conduit; the other is that mistakes under the slab are often more expensive to fix than mistakes overhead.

Slabbing is somewhat of an art; to do it well takes a little more care and concentration than installing conduit overhead. Here are several tips on slabbing:

1. Double-check the measurements. This is particularly critical when trying to hit the middle of an interior wall. The outside walls are easy to hit, because they are well-defined, but the inside walls are not often

well-marked before the concrete is poured, and sometimes the walls don't end up where they are supposed to be.

2. Always take your measurements from a permanent reference point, such as an outside wall and expansion joint or a floor joint.
3. Measure from more than one direction. When measuring to find an interior wall, measurements should be made from different points of view. Let's say that the first measurement is made from an outside wall, and the location is marked. Next, another measurement should be taken from some type of permanent reference in the other direction. Frequently, you will find that the building isn't built exactly as it was supposed to be. At this point, whoever is in charge of the construction of the building must be contacted to find out which way you should move the wall. Even though this is an extra step, it is far better than having to cut the slab to move your conduits.
4. You need to have workers on the job the day the concrete is poured. Concrete is very heavy and cumbersome to work with, so often the concrete finishers will knock something over in the process of spreading the concrete. Even if the finishers go out of their way to put whatever they pushed around back in place, their chances of getting it exactly correct are not too great. Have your workers wear boots like the finishers use and wade right in with them if necessary. Make sure that your installation is exactly as it is supposed to be. An hour later will be too late.
5. Make sure that all of your conduits are covered. Concrete is a lot easier to keep out of your pipes than it is to get out later. Pipes often get knocked down during a concrete pour. Duct tape is cheap and works well, but be sure that several layers are used so that the tape is not easily punctured.

6. Strap the conduits that go up an outside wall. If you can firmly strap these conduits and cover their open ends, you will almost never have any problems with them.
7. Make sure that the installation is double-checked. Have the construction superintendent check any locations that are questionable; you don't get many second chances with a concrete slab.

Prebending and Standardization

The technique of prebending and standardization can save a great amount of time in conduit installations. When properly used, it can enable an electrician to install many hundreds of feet of conduit per day.

The first step is to mount all of the boxes in the area where the installation is to go. They should all be mounted at preset standard heights (a certain preset height for every receptacle and a certain preset height for every switch, etc.). Now all of the conduit drops from the ceiling to either a switch or receptacle can be bent at one time. The steps to follow are these:

1. Verify the mounting heights (to the top of the box) of the boxes and the ceiling height and construction. You can then calculate exactly how long each type of conduit drop should be.
2. Place as many pieces of conduit as you will need so that one end of each conduit butts up against a wall. Make sure that all the pieces are even with each other. Now a line can be drawn across all of them at once, and they can all be bent to exactly the same length, whatever length it is that will be correct for the installation.
3. Cut or notch any of the framing members that will need to be cut or notched. Do this all at one time, rather than one-by-one each time the electrician comes to a new location.

4. Install all of the various drops. Have the electricians wear some type of work apron so that they will have all of the connectors, straps, and fasteners that they will need with them and won't have to spend time looking around for them.
5. Last, tie all of the conduits together overhead to complete the installation. If you go through the plans and mark which drops will be handled this way, it will simplify the matter quite a bit.

Trim Carts

Making special carts for carrying all necessary trim materials (switches, receptacles, plates, wire nuts, grounding pig-tails, flexible conduit, and fittings) can be a good time-saver. The savings here (as with most of these labor-saving techniques) is that the workers don't have to spend time trying to find things that they need—everything is right in front of them. The idea is to mimic the assembly-line process as much as possible. Without question, electrical installations are vastly different from building automobiles, but the more often you can arrange your labor force to work in this manner the more money you will save.

One other thing that can be done to save you a lot of labor on trim-out (if you can bring yourself to do it) is to get some little dollies. (A dolly is a piece of wood about 1-ft square with wheels on the bottom.) The electrician can sit on the dolly when wiring receptacles. By doing this, the worker doesn't have to get up, squat down, get up, and squat down; the worker can just scoot from outlet to outlet. One contractor who dared to try this said that he saved 40 percent of his trim-out labor by using the little scooters.

13. COMMUNICATIONS WIRING

The installation of communications wiring has become a significant part of the electrical trade in the past decade. Especially since the deregulation of phone companies, the installation of communications circuits has become the province of the electrician.

Common Telephone Connections

The most common and simplest type of communication installation is the single-line telephone. The typical telephone cable contains four wires, colored green, red, black, and yellow. A one-line telephone requires only two wires to operate. In almost all circumstances, green and red are the two conductors used. In a common four-wire modular connector, the red and green conductors are found in the inside positions, with the yellow and black wires on the outer positions.

As long as the two center conductors of the jack (again, always green and red) are connected to live phone lines, the telephone should operate.

Two-line phones generally use the same four-wire cables and jacks. In the case of two-line phones, however, the inside two wires (red and green) carry line 1, and the outside two wires (black and yellow) carry line 2.

Color Coding of Cables

The color coding of twisted-pair cable uses a color pattern that identifies not only which conductors make up a pair but also the place of the pair in the sequence relative to other pairs within a multipair sheath. This color coding is also used to determine which conductor in a pair is the "tip" conductor and which is the "ring" conductor. (The tip conductor is the positive conductor, and the ring conductor is the negative conductor.)

The banding scheme uses two opposing colors to represent a single pair. One color is considered the primary, whereas the other color is considered the secondary. For example, given the primary color of white and the secondary color of blue, a single twisted-pair would consist of one cable that is white with blue bands on it. The five primary colors are white, red, black, yellow, and violet.

In multipair cables the primary color is responsible for an entire group of pairs (five pairs total). For example, as shown in Fig. 13-1, the first five pairs all have the primary color of white. Each of the secondary colors—blue, orange, green, brown, and slate—are paired in a banded fashion with white. This continues through the entire primary color scheme for all five primary colors (comprising 25 individual pairs). In larger cables (50 pairs and up), each 25-pair group is wrapped in a pair of ribbons, again representing the groups of primary colors matched with their respective secondary colors. These color-coded band markings help cable technicians to quickly identify and properly terminate cable pairs.

Twisted-Pair Plugs and Jacks

One of the more important elements with respect to twisted-pair implementations is the cable jack or cross-connect block. These items are vital because without the proper interface any twisted-pair cable would be relatively useless. In the twisted-pair arena there are three major types of twisted-pair jacks:

- RJ-type connectors (phone plugs)
- Pin connectors
- Genderless connectors (IBM sexless data connectors)

The RJ-type (registered jack) name generally refers to the standard format used for most telephone jacks. The term pin connector refers to twisted-pair connectors, such as the RS-232 connector, which provide connection through male and female pin receptacles. Genderless connectors are connectors in which there is no separate male or female component; each component can plug into any other similar component.

Paint	*Tip (+) color*	*Ring (–) color*
1	white	blue
2	white	orange
3	white	green
4	white	brown
5	white	slate
6	red	blue
7	red	orange
8	red	green
9	red	brown
10	red	slate
11	black	blue
12	black	orange
13	black	green
14	black	brown
15	black	slate
16	yellow	blue
17	yellow	orange
18	yellow	green
19	yellow	brown
20	yellow	slate
21	violet	blue
22	violet	orange
23	violet	green
24	violet	brown
25	violet	slate

Fig. 13-1 Telecom color coding.

Standard Phone Jacks

The standard phone jack is specified by a variety of different names, such as RJ and RG, which refer to their physical and electrical characteristics. These jacks consist of a male and a female component. The male component snaps into the female receptacle. The important point to note, however, is the number of conductors each type of jack can support.

Common configurations for phone jacks include support for four, six, or eight conductors. A typical example of a four-conductor jack, supporting two twisted pairs, would be the one used for connecting most telephone handsets to their receivers.

A common six-conductor jack, supporting three twisted pairs, is the RJ-11 jack used to connect most telephones to the telephone company or PBX systems. An example of an eight-conductor jack is the R-45 jack, which is intended for use under the ISDN system as the user-site interface for ISDN terminals.

In the building-wiring field, the six-conductor jack is by far the most common. The eight-conductor jack, however, is rapidly becoming the most widely installed jack because more corporations are pulling twisted-pair in four-pair bundles for both voice and data. The eight-conductor jack, in addition to being used for ISDN, is also specified by several other popular applications, such as the new IEEE 802.3 10BaseT standard for Ethernet over twisted-pair.

These types of jacks are often keyed so that the wrong type of plug cannot be inserted into the jack. There are two kinds of keying—side keying and shift keying.

Side keying uses a piece of plastic that is extended to one side of the jack. This type is often used when multiple jacks are present.

Shift keying entails shifting the position of the snap connector to the left or right of the jack rather than leaving it in its usual center position. Shift keying is more commonly used for data connectors than for voice connectors.

Note that although these jacks are used for certain types of systems (data, voice, etc.), there is no official standard. Jacks can be used as you please. We only mention where they are used for informational purposes. They don't have to be used these ways.

Pin Connectors

There are any number of pin type connectors available. The most familiar type is the RS232 jack that is commonly used for computer ports. Another popular type of pin connector is the DB-type connector, which is the round connector that is commonly used for computer keyboards.

The various types of pin connectors can be used for terminating as few as five (the DB type) or more than 50 (the RS type) conductors.

The 50-pin *champ*-type connectors are often used with twisted-pair cables when connecting to cross-connect equipment, patch panels, and communications equipment, such as is used for networking.

Genderless Connectors

Genderless connectors are used almost exclusively for token-ring networks. Unlike the male and female types, the genderless connector is not differentiated as to male and female. Any genderless connector can plug together with any other genderless connector.

These connectors are generally known as *IBM data connectors* and are very seldom used except for IBM systems.

Cross Connects

Cross connections are made at terminal *blocks*. A block is typically a rectangular white plastic unit with metal connection points. The most common type is called a *punch-down* block. This is the kind that you see on the back wall of a business, where the main telephone connections are made. It is called a punch-down block because the wire connections are made by simply pushing the insulated wires into their places. When "punched" down, the connector cuts through the insulation and makes the appropriate connection.

Connections are made between punch-down blocks by using *patch cords*. Patch cords are simply short lengths of cable that can be terminated into the punch-down slots or that are equipped with connectors on each end.

When different systems must be connected, cross-connects are used.

Installation Requirements

Article 800 of the National Electrical Code (NEC) covers communication circuits such as telephone systems and outside wiring for fire and burglar alarm systems. Generally these circuits must be separated from power circuits and must be grounded. In addition, all such circuits that run outdoors (even if only partially) must be provided with circuit protectors (surge or voltage suppressors).

The requirements for these installations are as given in the following sections.

Conductors Entering Buildings

If communications and power conductors are supported by the same pole or run parallel in span, the following conditions must be met:

1. Wherever possible, communications conductors should be located below power conductors.
2. Communications conductors cannot be connected to crossarms.
3. Power service drops must be separated from communications service drops by at least 12 in.

Above roofs, communications conductors must have the following clearances:

1. Flat roofs: 8 ft.
2. Garages and other auxiliary buildings: None required.

3. Overhangs, where no more than 4 feet of communications cable will run over the area: 18 in.
4. Where the roof slope is 4-in. rise for every 12 in. horizontally: 3 ft.

Underground communications conductors must be separated from power conductors in manholes or handholes by brick, concrete, or tile partitions.

Communications conductors should be kept at least 6 ft away from lightning protection system conductors.

Circuit Protection

Protectors are surge arresters designed for the specific requirements of communications circuits. They are required for all aerial circuits not confined within a block. (*Block* here means city block.) They must be installed on all circuits within a block that could accidentally contact power circuits over 300 V to ground. They must also be listed for the type of installation.

Other requirements are the following:

- Metal sheaths of any communications cables must be grounded or interrupted with an insulating joint as close as practicable to the point where they enter any building (such point of entrance being the place where the communications cable emerges through an exterior wall or concrete floor slab or from a grounded rigid or intermediate metal conduit).
- Grounding conductors for communications circuits must be copper or some other corrosion-resistant material and must have insulation suitable for the area in which the conductors are installed.
- Communications grounding conductors may be no smaller than No. 14.
- The grounding conductor must be run as directly as possible to the grounding electrode and must be protected if necessary.

- If the grounding conductor is protected by metal raceway, the raceway must be bonded to the grounding conductor on both ends.

Grounding electrodes for communications ground may be any of the following:

- The grounding electrode of an electrical power system.
- A grounded interior metal piping system. (Avoid gas piping systems for obvious reasons.)
- Metal power service raceway.
- Power service equipment enclosures.
- A separate grounding electrode.

If the building being served has no grounding electrode system, the following can be used as a grounding electrode:

1. Any acceptable power system grounding electrode (see *Section 250.52* of the NEC).
2. A grounded metal structure.
3. A ground rod or pipe at least 5 ft long and ½ in. in diameter. This rod should be driven into damp (if possible) earth and should be kept separate from any lightning protection system grounds or conductors.

Connections to grounding electrodes must be made with approved means.

If the power and communications systems use separate grounding electrodes, they must be bonded together with a No. 6 copper conductor. Other electrodes may be bonded also. This is not required for mobile homes.

For mobile homes, if there is no service equipment or disconnect within 30 ft of the mobile home wall, the communications circuit must have its own grounding electrode. In this case, or if the mobile home is connected with cord and plug, the communications circuit protector must be bonded to the mobile home frame or grounding terminal with a copper conductor no smaller than No. 12.

Interior Communications Conductors

Communications conductors must be kept at least 2 in. away from power or Class 1 conductors, unless they are permanently separated from each other or unless the power or Class 1 conductors are enclosed in one of the following:

1. Raceway
2. Type AC, MC, UF, NM, or NM cable, or metal-sheathed cable

Communications cables are allowed in the same raceway, box, or cable with any of the following:

1. Class 2 and 3 remote-control, signaling, and power-limited circuits
2. Power-limited fire protective signaling systems
3. Conductive or nonconductive optical fiber cables
4. Community antenna television and radio distribution systems

Communications conductors are not allowed to be in the same raceway or fitting with power or Class 1 circuits.

Communications conductors are not allowed to be supported by a raceway unless the raceway runs directly to the piece of equipment the communications circuit serves.

Openings through fire-resistant floors, walls, and so on must be sealed with an appropriate fire-stopping material.

Any communications cables used in plenums or environmental air-handling spaces must be listed for such use.

Communications and multipurpose cables can be installed in cable trays.

Any communications cables used in risers must be listed for such use.

Cable substitution types are shown in *Table 800.53* of the NEC.

Structured Cabling

Modern data networks are typically called structured cabling or Ethernet. Ethernet was one of the first types of networks. Structured cabling is a complete system of cabling and associated hardware and must conform to the EIA/TIA 568 Structured Cabling Standard.

Structured cabling generally refers to a network cabling system that is designed and installed according to preset standards. The benefits of structured cabling are the following:

- Buildings, new or refurbished, are prewired without needing to know future occupant's data communication needs.
- Future growth and reconfiguration accommodated by predefined topologies and physical specifications, such as distances.
- Support of multivendor products, including cables, connectors, jacks, plugs, adapters, baluns, and patch panels.
- Voice, video, and all other data transmissions are integrated.
- Cable plants are easily managed and faults are readily isolated.
- All data cabling work can be accomplished while other building work is taking place.

EIA/TIA 568

By far the most commonly used standard for structured wiring systems is EIA/TIA 568, or 568A. The EIA/TIA 568 wiring standard recognizes four cable types and two types of telecommunications outlets.

Some of the parameters of 568 are the following:

- Up to 50,000 users
- 90-meter horizontal distance limit between closet and desktop
- 4 pairs of conductors to each outlet—all must be terminated
- 25-pair cables may *not* be used (crosstalk problems)
- May not use old wiring already in place
- Bridge taps and standard telephone wiring schemes may not be used
- Requires extensive testing procedures

Following are the standard materials used for a structured cabling (568) system:

- Four-pair 100 ohm UTP cables. The cable consists of 24 AWG thermoplastic insulated conductors formed into four individually twisted pairs and enclosed by a thermoplastic jacket. Four-pair, 22 AWG cables that meet the transmission requirements may also be used. Four-pair, *shielded* twisted-pair cables that meet the transmission requirements may also be used.
- The diameter over the insulation shall be 1.22 mm (0.048 in) max.
- The pair twists of any pair shall not be exactly the same as any other pair. The pair-twist lengths shall be selected by the manufacturer to ensure compliance with the crosstalk requirements of this standard.

Transmission Problems

Sending data through copper conductors is essentially the same as sending power through copper conductors except that the amounts of current and the conductors are far smaller and the voltage and current characteristics differ.

For power work, you are concerned about the path the current will take, but you have fairly little concern for the quality of the power going from one point to another. For data wiring, you must consider two qualities of the transmission:

1. **There must be a clear path from one machine to the next.** Here you are concerned with the signal's strength; it must arrive at the far end of the line with enough strength to be useful.
2. **The signal must be of good quality.** For instance, if you send a square-wave digital signal into one end of a cable, you want a good square wave coming out of the far end. If this signal is distorted, it is unusable, even if it is still strong.

The main problem you have with power wiring is a loss of power, which you call *voltage drop*. You have the same problem with data signals when you attempt to send them through conductors with too much resistance (usually due to distance). With data cabling, you call this *attenuation*, and it is virtually the same as voltage drop—not enough power is getting through. The problem of signals quality is a completely different concern. Getting enough signal from one end of the cable to another is one thing you must do; but you must also make sure that the signal at the far end is not too distorted to use.

Cable Installation

During a datacom installation, cables will be roughed in much the same as is done for common electrical cables. It is critical that manufacturer's instructions be followed precisely. Also important is the use of the proper connectors and fittings. There are many different types of such fittings, and the placement of the appropriate fittings is crucial to the final equipment connections.

When pulling main runs of cables, additional conditions must be met. In general, the installation of these conductors is accomplished by the same methods as with standard electrical wires, but extra care is necessary. Because the cable is made with far less copper, the tensile strength of the cable is proportionally reduced. Sidewall pressure also carries the risk of damaging the insulation of the individual conductors, which puts the performance of the product at risk. Some general rules follow:

1. Don't exceed a pulling tension of 20 percent of the ultimate breaking strength of the cable (these figures are available from the cable maker).
2. Lubricate the raceway generously with a suitable pulling compound. (Check with the manufacturer for types of lubricants that are best suited to the type of cable.)
3. Use pulling eyes for manhole installations.
4. For long underground runs, pull the cable both ways from a centrally located manhole to avoid splicing. Use pulling eyes on each end.
5. Do not bend, install, or rack any cable in a radius of less than 12 times the cable diameter.

Testing

Datacom installers are generally required to prove the quality of their materials and work by *certifying* all installed cable runs using equipment that is specifically designed for testing these data links. Certification simply means testing and documenting that the tests were passed. Digital multimeters alone are inadequate for this task. Although a good multimeter may be able to confirm that signal is getting through from one end of the cable to another, they cannot confirm the quality of the signal. Are their crisp pulses coming through? What is the ratio of attenuation to crosstalk? Unless your tester can do these things well, it is inadequate for data testing.

In addition, it is helpful to have a tester that saves its results and is set up to transfer them to a computer program for documentation. This feature will save a lot of time when completing the project.

Data cabling should be tested each time you install, move, or troubleshoot a LAN-attached workstation so that you can prevent cabling problems from impacting the performance of your high-speed network.

Understanding Decibels

In data cabling, most energy and power levels, losses or attenuations, are expressed in decibel rather than in the watt. The reason is simple. Transmission calculations and measurements are almost always made as *comparisons* against a reference: received power compared to emitted power, energy in versus energy out (energy lost in a connection), and so on.

Generally, energy levels (emission, reception, etc.) are expressed in dBm. This signifies that the reference level of 0 dBm corresponds to 1 milliwatt (mW) of power.

Generally, power losses or gains (attenuation in a cable, loss in a connector, etc.) are expressed in dB. The unit dB is used for very low levels.

Decibel measurement works as follows: A difference of 3 dB equals a doubling or halving of power.

A 3 dB gain in power means that the optical power has been doubled. A 6 dB gain means that the power has been doubled, and doubled again, equaling four times the original power. A 3 dB loss of power means that the power has been cut in half. A 6 dB loss means that the power has been cut in half, then cut in half again, equaling one fourth of the original power.

A loss of 3 dB in power is equivalent to a 50 percent loss, for example, 1 mW of power in and .5 mW of power out.

A 6 dB loss would equal a 75 percent loss (1 mW in, .25 mW out).

When To Test

The testing of network cables should be done both during the installation process and upon completion of the system. Testing during the installation process helps catch problems while they are still simple to fix. Testing the system upon completion is not only a good practice but is even required by law for communications systems.

Common Cable Test Equipment

The most common testing tools for copper data cabling are the following (optical fiber will be covered later in this lesson):

DVM (Digital Volt Meter). Measures volts.

DMM (Digital Multimeter). Measures volts, ohm, capacitance, and some measure frequency.

TDR (Time Domain Reflectometer). Measures cable lengths, locates impedance mismatches.

Tone Generator and Inductive Amplifier. Used to trace cable pairs, follow cables hidden in walls or ceiling. The tone generator will typically put a 2 kHz audio tone on the cable under test, the inductive amp detects and plays this through a built-in speaker.

Wiremap Tester. Checks a cable for open or short circuits, reversed pairs, crossed pairs, and split pairs.

Noise testers, 10BaseT. The standard sets limits for how often noise events can occur, and their size, in several frequency ranges. Various handheld cable testers are able to perform these tests.

Butt sets. A telephone handset that when placed in series with a battery (such as the one in a tone generator), allows voice communication over a copper cable pair. This can be used for temporary phone service in a wiring closet.

Cable Colors

Although the color coding of data cabling is not as well known (or followed) as the color code for power conductors, color codes do exist and should be followed. (In power wiring, things explode if you don't use the color code. In data cabling, they simply don't work.)

The Cabling Administration Standard (EIA 606) lists the colors and functions of data cabling as the following:

Blue	Horizontal voice cables.
Brown	Interbuilding backbone.
Gray	Second-level backbone.
Green	Network connections and auxiliary circuits.
Orange	Demarcation point, telephone cable from central office.
Purple	First-level backbone.
Red	Key-type telephone systems.
Silver or White	Horizontal data cables, computer, and PBX equipment.
Yellow	Auxiliary, maintenance, and security alarms.

Fiber Optic Work

There is no more appropriate technology for the massive advances in the communications fields than optical fiber cable. This communications medium (properly called optical fiber cables) can carry a much greater variety of communications signals and can carry them far more efficiently than other means.

Optical fiber cables are commonly being used for long-distance telephone communications and computer networks, and they are beginning to be used for residential telephone systems.

You can be sure that the numbers of these cables being installed will skyrocket in coming years.

Splicing Optical Fibers

The greatest part of labor on many optical fiber installations is involved with splicing and terminating the fibers. This is a long, tedious, and sometimes frustrating process. Often, times of up to 1 hour can be required *per splice*. Newer types of splices and terminations are continually improving and making these times decrease, yet splice time is still very significant.

As you install optical cable, be careful to avoid unnecessary splices. Splices are labor intensive, and too much splicing causes labor overruns.

Two kinds of splices are commonly used: fusion splices and mechanical splices. These splices can reduce losses to less than 0.2 dB.

Fusion splices are made by first cleaving the cable ends so that they are square. These ends are then aligned, and a short, controlled electrical arc melts the cable ends, fusing them into a continuous fiber. After the splice is complete, the splice is covered and supported to avoid breakage. Fusion splicing machines are quite expensive, and training must be fairly lengthy.

Mechanical splicing has been developed as an alternative to fusion splicing because fusion splicing cannot be done in all environments and because the cost per splice is high. *Mechanical splices* use the same principles as fusion splices, but the implementation is different. Losses are reduced by positioning the ends closely in a groove, elastomeric sleeve, adjustable bushing, or microtube. The ends are connected with either an index-matching gel or ultraviolet hardened optical adhesive. Mechanical splices have losses of 0.4 dB or less and can be done in more severe environments.

Cable Types

The physical construction of optical cables is not governed by any agency. It is up to the designer of the system to make sure that the cable selected will meet the application requirements. Four basic cable types have, however, emerged as de facto standards for a variety of applications:

Simplex and zip cord. One or two fibers, tight-buffered, Kevlar reinforced and jacketed. Used mostly for patch cord and backplane applications.

Tightpack cables. Also known as distribution-style cables. Up to several tight-buffered fibers bundled under the same jacket with Kevlar reinforcement. Used for short, dry conduit runs and for riser and plenum applications. These cables are small in size, but because their fibers are not individually reinforced, these cables need to be terminated inside a patch panel or junction box.

Breakout cables. These cables are made of several simplex units cabled together. This is a strong, rugged design and is larger and more expensive than the tight-pack cables. It is suitable for conduit runs and for riser and plenum applications. Because each fiber is individually reinforced, this design allows for a strong termination to connectors and can be brought directly to a *computer* backplane.

Loose tube cables. These are composed of several fibers cabled together to provide a small, high fiber-count cable. This type of cable is ideal for outside plant trunking applications. Depending on the actual construction, it can be used in conduits, strung overhead, or buried directly into the ground.

Hybrid or composite cables. There is a lot of confusion over these terms, especially since the NEC just switched its terminology from *hybrid* to *composite.* Under the new terminology, a *composite* cable is one that contains a number of copper conductors properly jacketed and sheathed, depending on the application, in the same cable assembly as the optical fibers. In issues of the code previous to 1993, this was called *hybrid* cable.

This terminology situation is made all the more confusing because there is another type of cable that is called composite.

This type of cable contains only optical fibers, but they are two different types of fibers: multimode and single-mode.

Remember that there is a great deal of confusion over these terms, with many people using them interchangeably. It is my contention that you should now use the term *composite* for fiber/copper cables since that is how they are identified in the NEC. Also, you should probably use *hybrid* for fiber/fiber cables, since the code gives us little choice.

Installation

Although the installation methods for both electronic wire cables and optical fiber cables are similar, there are two very important additional considerations that must be applied to optical fiber cables:

1. Never pull the fiber itself.
2. Never allow bends, kinks, or tight loops.

In order to keep these two rules, you must identify the strength member and fiber locations within the cables, then use the method of attachment that pulls most directly on the strength member. By paying careful attention to the strength limits and minimum bending radius limits, and by avoiding scraping at sharp edges, damage can be avoided.

Cables in Trays

Optical fiber cables in trays should be carefully placed without tugging on the outer jacket of the cable. Care must be taken so that the cables are placed where they cannot be crushed. Flame-retardant cables are recommended for interior installations.

Vertical Installation

Optical fibers in any type of vertical tray, raceway, or shaft should be clamped at frequent intervals so that the entire weight of the cable is supported at the top. The weight of the cable should be evenly supported over its entire length. Clamping intervals may vary from between 3 ft for outdoor installations with wind stress problems to 50 ft for indoor installations.

When installed vertically, the fibers sometimes have a tendency to migrate downward, especially in cold weather, which causes a signal loss (attenuation). This migration can be prevented by placing several loops of about 1 ft in diameter at the top of the run, at the bottom of the run, and at least once every 500 ft in between.

Cables in Conduit

For all but the shortest pulls, loose buffer cables are preferred because they are stiffer and their jackets generally cause less friction than tight buffer cables.

The cable lubricant must be matched to the jacket material of the cable. Most commercial lubricants will be compatible with popular types of cable jackets, but not in every case. Lubrication is considerably more important for optical fiber cables than for copper cables because the fibers can be easily damaged.

In difficult installations, the cable-pulling force should be monitored with a tension meter. In these cases, the conduit should be prelubricated and the cable lubricated also as it is installed. Special lubricant spreaders and applicators are often used as well.

Except when tension meters are used, cable pulling should be done by hand and should be in continuous pulls as much as possible. Often this means pulling from a central manhole or pullbox. During the pulling process, all tight bends, kinks, and twists must be carefully avoided. If they are not, the damaged cable may need to be removed and replaced with undamaged cable.

Attachment

The proper method of pulling optical fiber cables is to attach the pull wire or tape to the cable's strength member. This avoids any tension on the fibers themselves. Unfortunately it is not always easy to do.

When attaching to the strength members, the outer coverings are stripped back. Care must be taken not to damage the strength members, but attachment can normally be done with common tools. Kevlar or steel strength members can be tied directly to the pulling eye. Other more rigid types of strength members (such as fiberglass-epoxy) must be connected to a special set-screw device.

Indirect attachment can usually be done well with Kellems grips that firmly grip the cable jacket. For some larger cables, this type of attachment can actually be preferred. If you prestretch the Kellems grip and tape it firmly to the cable, much of the cable strain will be avoided.

Indirect attachment is not desirable when the fibers will be in the path of the forces between the pulling grip and the strength members. This is the case when the strength member is in the center of the cable, surrounded by the fibers. In such cases, only a small pulling force can be used.

Direct Burial

Generally, only heavy-duty cables can be directly buried. There are numerous hazards that affect directly buried optical fiber cables, such as freezing water, rocky soils, construction activities, and rodents (usually gophers). Burying the cables at least 3 or 4 ft deep avoids most of these hazards, but only strong metal braids or cables too large to bite will deter the gophers.

When plowing is used as an installation means, only loose buffer cables are used, because they can withstand uneven pulling pressures better than tight buffer cables. Where freezing water presents a problem, metal sheaths, double jackets, and gel fillings can be used as water barriers.

Rather than using expensive, heavy-duty cables, 1-in. polyethylene gas pipe is sometimes used to form a simple conduit. These tubes are also used as inner ducts, placed inside of larger (usually 4-in.) conduits. The plastic pipes provide a smooth passageway; and by using several units inside of the larger conduit (with spacers holding them in place), the cables stay well organized. The plastic pipe can be smoothly bent, providing for very convenient installations.

Aerial

When optical fibers are to be installed aerially, they must be supported by a messenger wire. (See *Article 396* of the NEC.) Round, loose buffer cables are preferred; they should be firmly and frequently clamped to the messenger wire.

Cables for long outdoor runs are usually temperature stabilized. Steel is used for the stabilization if there are no lightning or electrical hazards. In other cases, fiberglass-epoxy is used. This type of all-dielectric cable is preferred for high vertical installations, such as TV or radio towers.

Tools

There are certain items that are absolutely necessary for fiber work. You must have the right equipment to get the fibers in place, terminate them, and test them. No matter how you start, there is no way around purchasing the following items:

- Power meter
- Swivel pulling eyes
- Breakaway swivel or tension meter
- Microscopes
- Polishing "pucks"
- Termination kits
- Sandpaper
- Adhesive syringes
- Cleavers
- Stripping tools
- Solvent and wipes
- Canned air
- Adhesives

Fiber Requirements

Remember that the NEC designates cable types differently from the way the rest of the trade does. The code specifies horizontal cables, riser-rated cables, and plenum-rated cables. It also specifies cables as *conductive* or *nonconductive*. Note that a conductive cable is a cable that has any metal in it at all. The metal in a conductive cable does not have to be used to carry current; it may simply be a strength member.

The main requirements of *Article 770* are the following:

When optical cables that have *non*current-carrying conductive members contact power conductors, the conductive member must be grounded as close as possible to the point at which the cable enters the building. If desired, the conductive member may instead be broken (with an insulating joint) near its entrance to the building.

Nonconductive optical cables can share the same raceway or cable tray as other conductors operating at up to 600 V.

Composite optical cables can share the same raceway or cable tray as other conductors operating at up to 600 V.

Nonconductive optical cables cannot occupy the same enclosure as power conductors, except in the following circumstances:

1. When the fibers are associated with the other conductors.
2. When the fibers are installed in a factory-assembled or field-assembled control center.
3. Nonconductive optical cables or hybrid cables can be installed with circuits exceeding 600 V in industrial establishments, where they will be supervised only by qualified persons.

Both conductive and nonconductive optical cables can be installed in the same raceway, cable tray, or enclosure with any of the following:

- Class 2 or 3 circuits.
- Power-limited fire protective signaling circuits.
- Communication circuits.
- Community antenna television circuits.
- Composite cables must be used exactly as listed on their cable jackets.
- All optical cables must be installed according to their listings.

Refer to *Table 770.53* of the NEC to see the cable substitution hierarchy.

14. WIRING IN HAZARDOUS LOCATIONS

Wiring in hazardous locations is subject to much more stringent requirements than other types of wiring. These locations come with serious built-in dangers. What might be a minor problem in an office environment could be life-threatening in a propane storage terminal.

Virtually all hazardous locations are covered in *Articles 500* through *517* of the NEC. These articles should be well understood by anyone wiring such locations. These areas are dangerous for a number of reasons, and each has its own special requirements and hazards.

It is also important to remember that wiring in hazardous locations is very expensive. The special types of equipment (explosion-proof equipment, for example) cost many times more than standard types, and the amount of labor necessary to install them is very high. Therefore, it is important to use ingenuity in laying out wiring for hazardous locations. Your goal is to install as much of the wiring system as possible outside of the hazardous areas.

Remember that all electrical installations in hazardous locations are inherently dangerous. Don't perform installations without carefully engineered layouts. If you don't have first-rate instructions, don't install the wiring! The installation requirements in this chapter are given to assist in the installation process, not to substitute for an engineered layout. This work can be dangerous—don't take chances.

The following requirements are found in *Articles 500* through *505* of the NEC:

General Considerations

There are five primary requirements for hazardous locations:

1. Locations are classified as hazardous depending on the nature of the chemicals, dust, or fibers that may be

present and also on their concentrations in the various environments.

2. In determining classifications, each room or area is considered separately.
3. All equipment installed in hazardous areas (also called classified areas) must be approved for the specific area in which it is installed, not just approved for hazardous locations in general.
4. The wiring requirements for one type of hazardous location cannot be substituted for the requirements of another location. They are not interchangeable.
5. Locknut-bushing or double-locknut connections are not considered adequate bonding methods for hazardous locations. Other methods must be used.

Class I Locations

Class 1 locations are areas where flammable gases or vapors are present in amounts great enough to produce explosive or ignitable mixtures.

Class 1, Division 1 locations are areas where flammable concentrations of gases or vapors may be present under normal operating conditions; or where such gases are frequently present because of maintenance or leakage; or where a breakdown might cause such vapors or gases to be present. One such location is the area around dispensing valves for propane or other flammable gases.

Class 1, Division 2 locations are areas where flammable liquids, vapors, or gases are handled or processed, but where the liquids, vapors, or gases are normally contained in closed containers from which they will escape only in abnormal cases. Typical of these locations is the area around propane storage tanks.

Wiring methods in Class 1 groups are determined by the ignition temperatures for the types of gases or vapors in them.

Wiring methods in Class 1, Division 1 areas may be any of the following:

1. Threaded rigid metal conduit.
2. Threaded intermediate metal conduit.
3. Type MI cable (using fittings suitable for the location).

All boxes, fittings, and joints must be threaded for conduit and cable connections and must be explosion-proof.

Threaded joints must be made up with at least five threads fully engaged.

Type MI cable must be carefully installed so that no tensile (pulling) force is placed on the cable connectors.

Flexible connections can be used only where necessary and must be made with materials approved for Class 1 locations.

Seals must be provided within 18 in. of any enclosure housing equipment that can produce sparks, such as switches, relays, and circuit breakers, for example (see Figs. 14-1 and 14-2).

Seals must also be provided where conduits enter or leave a Class 1, Division 1 area. The seals may be on either side of the dividing wall (see Figs. 14-3 and 14-4).

Wiring methods in Class 1, Division 2 areas may be any of the following:

1. Threaded rigid metal conduit.
2. Threaded intermediate metal conduit.
3. Type MI, MV, MC, TC, or SNM cable (using fittings suitable for the location). These cables must be carefully installed so that no tensile (pulling) force is placed on the cable connectors.
4. Enclosed gasketed busway.
5. Enclosed gasketed wireway.
6. Type PLTC cable (installed under the provisions of *Article 725* of the NEC).
7. Type MI, MV, MC, TC, SNM, and PLTC cables may be installed in cable trays.

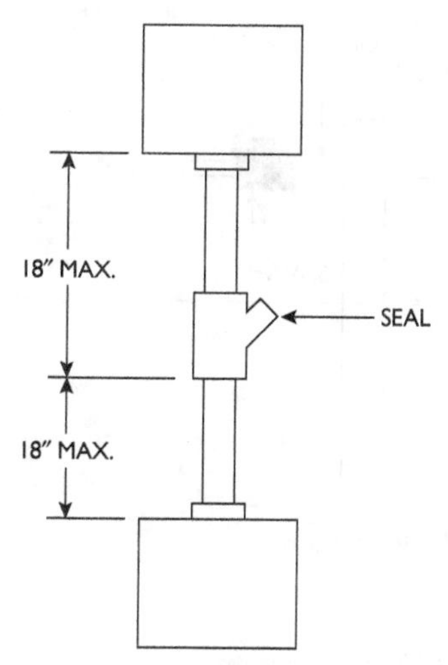

Fig. 14-1 Seal placement on a short run.

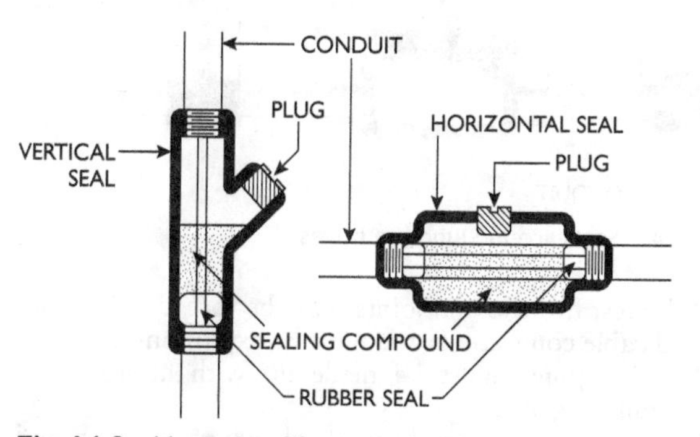

Fig. 14-2 Vertical and horizontal seals.

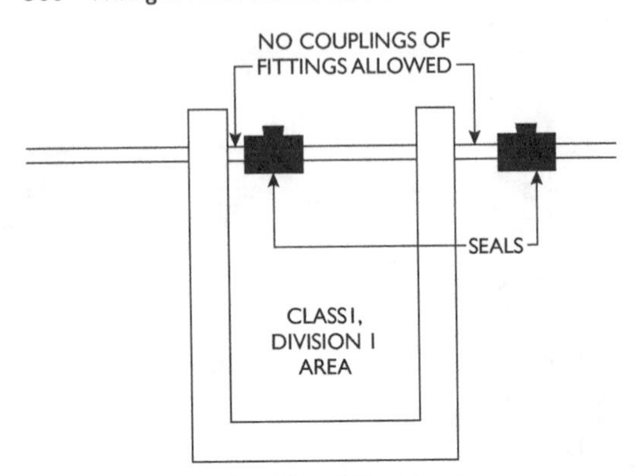

Fig. 14-3 Seals for Class I Division I area.

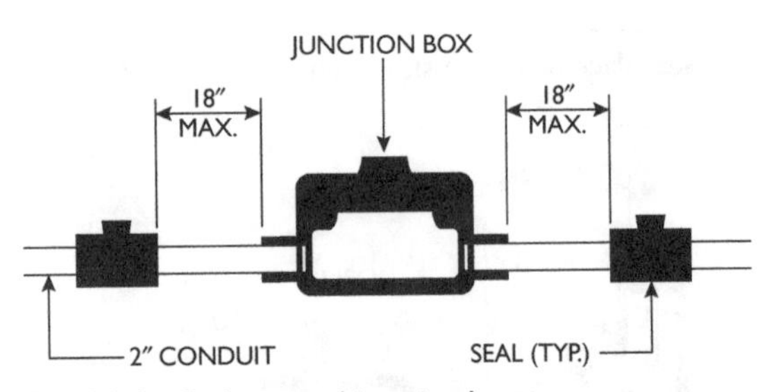

Fig. 14-4 Seals around junction boxes.

All boxes, fittings, and joints must be threaded for conduit and cable connections and must be explosion-proof.

Threaded joints must be made up with at least five threads fully engaged.

Flexible connections can be used only where necessary and must be made with materials approved for Class 1 locations.

Class 2 Locations

Class 2 locations are areas in which combustible dust may be present.

Class 2, Division 1 locations are areas where the concentrations of flammable dust under normal conditions are sufficient to produce explosive or ignitable mixtures.

Class 2, Division 2 locations are areas where flammable or ignitable dust is present, but normally not in quantities sufficient to produce a flammable or explosive mixture.

Wiring methods in Class 2, Division 1 areas may be any of the following:

1. Threaded rigid metal conduit.
2. Threaded intermediate metal conduit.
3. Type MI cable (using fittings suitable for the location).

All boxes, fittings, and joints must be threaded for conduit and cable connections and must be approved for Class 2 locations (see Fig. 14-5).

Type MI cable must be carefully installed so that no tensile (pulling) force is placed on the cable connectors.

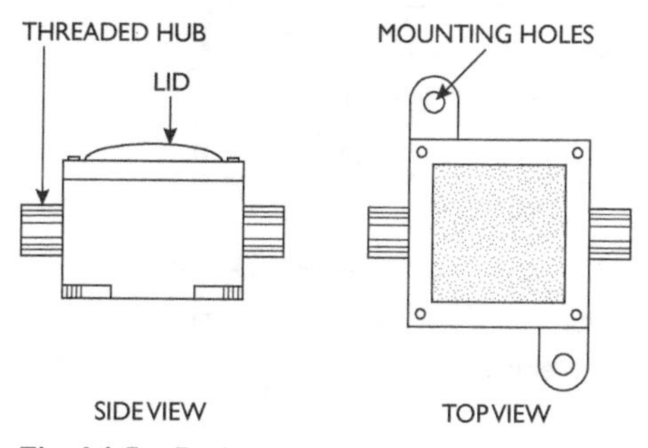

Fig. 14-5 Explosion-proof box.

Flexible connections can be made with any of the following:

1. Liquid-tight flexible metal conduit (with approved fittings).
2. Liquid-tight flexible nonmetallic conduit (with approved fittings).
3. Extra-hard-usage cord, with bushed fittings and dust seals.

When raceways extend between Class 2, Division 1 locations, sealing may be done in any of the following ways (see Fig. 14-6):

1. With raceway seals.
2. With a 10-ft horizontal run of raceway.
3. With a 5-ft vertical raceway.

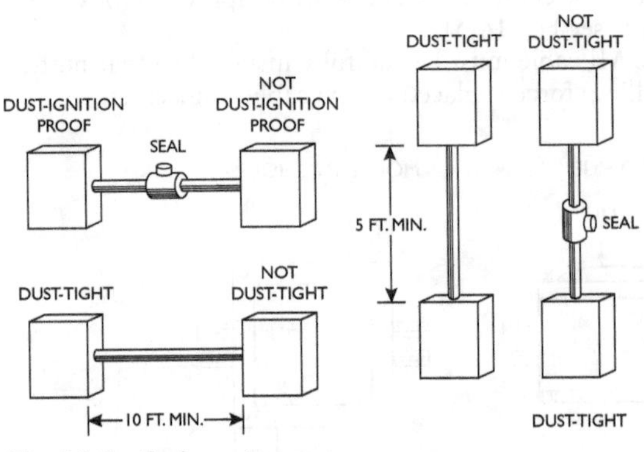

Fig. 14-6 Seal requirements.

Wiring methods in Class 2, Division 2 locations may be any of the following:

1. Threaded rigid metal conduit.
2. Threaded intermediate metal conduit.
3. Type MI, MC, or SNM cable (using fittings suitable for the location). These cables must be carefully installed so that no tensile (pulling) force is placed on the cable connectors.
4. Enclosed gasketed busway.
5. Enclosed gasketed wireway.
6. Types MC, TC, and PLTC cables may be installed in cable trays.

When raceways extend between Class 2, Division 2 locations and unclassified locations, sealing may be done in any of the following ways:

1. With raceway seals.
2. With a 10-ft horizontal run of raceway.
3. With a 5-ft vertical raceway.

Flexible connections can be made by any of the following means:

1. Liquid-tight flexible metal conduit (with approved fittings).
2. Liquid-tight nonmetallic conduit (with approved fittings).
3. Extra-hard-usage cord, with bushed fittings and dust seals.

Class 3 Locations

Class 3 locations are termed hazardous areas only because of the presence of easily ignitable fibers or flyings, but these fibers or flyings are not likely to be suspended in the air in quantities sufficient to cause ignitable mixtures.

Class 3, Division 1 locations are areas where easily ignitable fibers or materials producing easily ignitable flyings are handled, manufactured, or used.

Class 3, Division 2 locations are areas where ignitable fibers are stored or handled.

Wiring methods in Class 3, Division 1 or 2 locations may be any of the following:

1. Threaded rigid metal conduit.
2. Threaded intermediate metal conduit.
3. Type MI, MC, SNM cable (using fittings suitable for the location). These cables must be carefully installed so that no tensile (pulling) force is placed on the cable connectors.
4. Dust-tight wireway.

Flexible connections can be made by any of the following means:

1. Liquid-tight flexible metal conduit (with approved fittings).
2. Liquid-tight flexible nonmetallic conduit (with approved fittings).
3. Extra-hard-usage cord with bushed fittings and dust seal.

Intrinsically Safe Systems

Intrinsically safe systems must be designed for the specific installation, complete with control drawings. This is required by the NEC, and no intrinsically safe system installation may be attempted without such drawings.

Commercial Garages

Commercial garages are classified as hazardous areas, although not all areas of a particular garage may be considered hazardous. The requirements for commercial garages are shown in *Article 511* of the NEC, as follows:

In commercial garages, the entire area of the garage, from the floor up to a height of 18 in., must be considered a Class 1, Division 2 location (see Fig. 14-7).

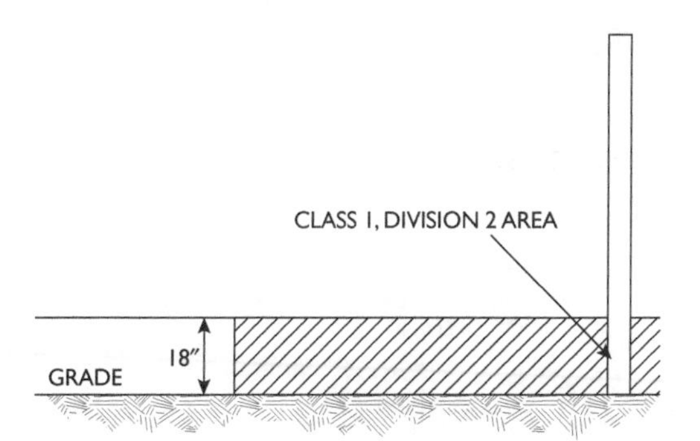

Fig. 14-7 Class 1, Division 2 area.

Any pit or depression in the floor shall be considered a Class 1, Division 1 location from the floor level down (see Figs. 14-8 and 14-9).

Adjacent areas such as storerooms, switchboard rooms, etc., are not considered to be classified areas if they have ventilating systems that provide four or more air changes per hour, or if they are well separated from the garage area by walls or partitions.

Areas around fuel pumps are covered by *Article 514* of the NEC.

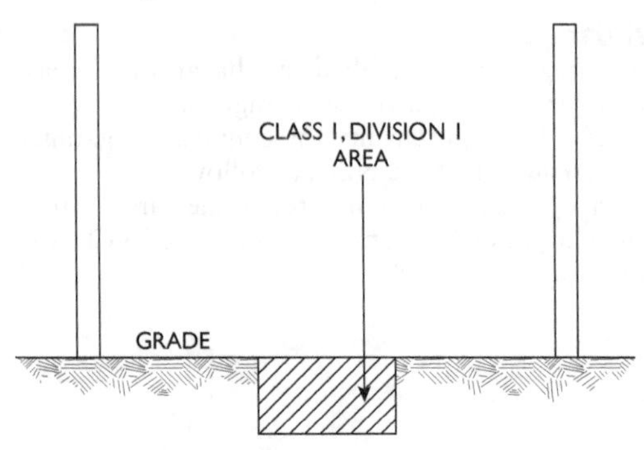

Fig. 14-8 Class 1, Division 1 area.

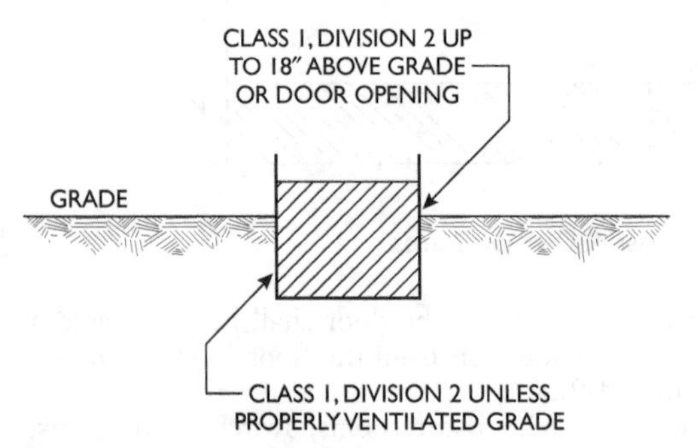

Fig. 14-9 Class 1 location.

Wiring methods in areas above Class 1 locations may be any of the following:

1. Rigid metal conduit
2. Intermediate metal conduit
3. Rigid nonmetallic conduit

4. Electrical metallic tubing
5. Type MI, TC, SNM, or MC cable

Plug receptacles above Class 1 locations must be approved for the purpose.

Electrical equipment that could cause sparks and is located above Class 1 locations must be totally enclosed or located at least 12 feet from the floor.

All receptacles installed where hand tools, diagnostic equipment, or portable lighting devices are to be used must have ground-fault protection.

Aircraft Hangars

The requirements for aircraft hangars are shown in *Article 513* of the Code. It is important that in addition to concerns over volatile fuels, special care must also be taken with aircraft because of problems with static electricity. Static can cause sparks that, in turn, can ignite volatile substances such as gasoline. This requires special attention to grounding systems in these areas.

The requirements of *Article 513* are as follows:

In aircraft hangars, the entire area from the floor up to a height of 18 in. must be considered a Class 1, Division 2 location (see Fig. 14-10).

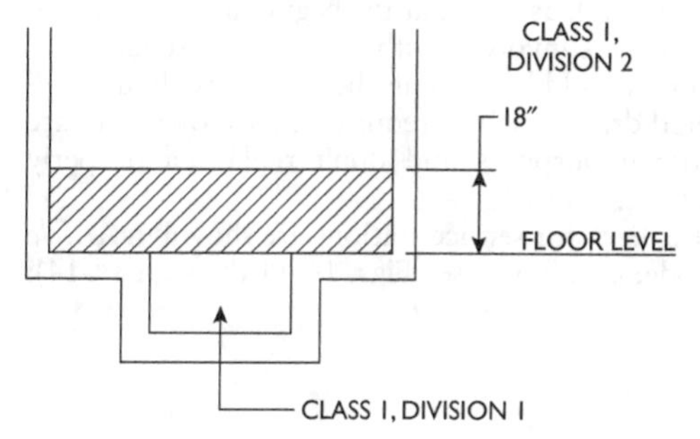

Fig. 14-10 Class I location.

Any pit or depression in the floor shall be considered a Class 1, Division 1 location from the floor level down.

Adjacent areas such as storerooms and switchboard rooms are not considered to be classified areas if they are well separated from the hangar itself by walls or partitions.

Wiring methods in areas above Class 1 locations may be any of the following:

1. Rigid metal conduit
2. Intermediate metal conduit
3. Electrical metallic tubing
4. Type MI, TC, SNM, or MC cable

Electrical equipment that could cause sparks and is located above Class 1 locations must be totally enclosed or located at least 10 feet from the floor.

Aircraft electrical systems must be deenergized when an aircraft is stored in a hangar and, whenever possible, when the aircraft is being serviced.

Aircraft batteries cannot be charged when they are installed in an aircraft located partially or fully inside a hangar.

Service Stations

Gasoline dispensing facilities and service stations are hazardous locations, and, as stated at the beginning of this chapter, there are dangers involved with wiring in these facilities. No installations should be done in these areas without carefully engineered drawings. The requirements shown here are for informational purposes and don't replace a properly engineered layout.

The requirements for service stations are shown in *Article 514* of the Code, as follows (see Figs. 14-11 through 14-14):

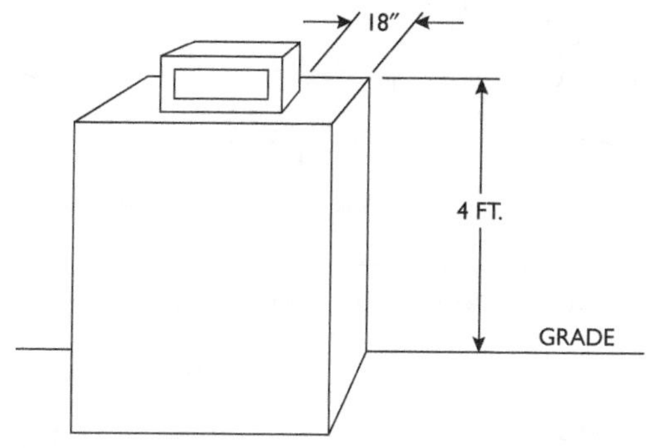

Fig. 14-11 Class I locations around dispensing stations.

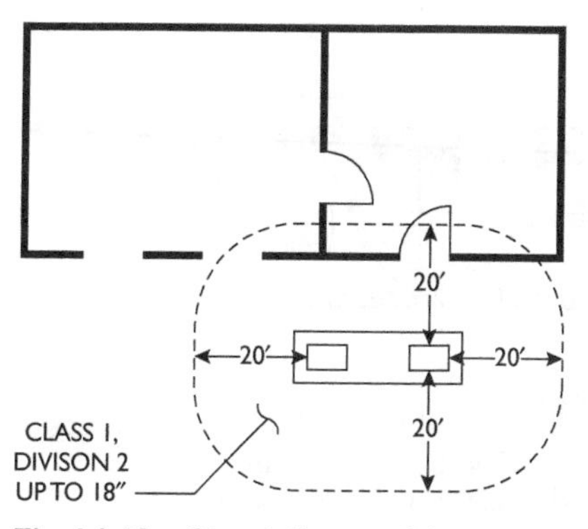

Fig. 14-12 Class 1, Division 2 locations.

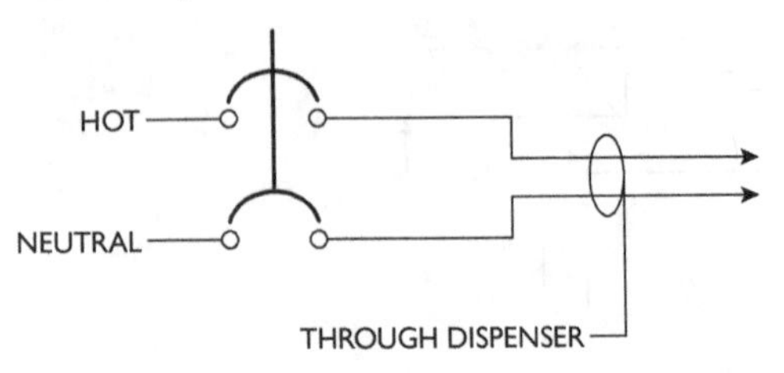

Fig. 14-13 Circuit breaker for dispensing locations.

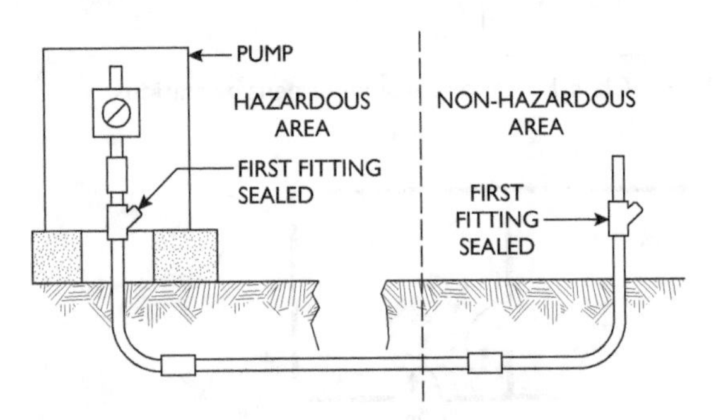

Fig. 14-14 Definitions of hazardous areas.

Installation Methods

Table 514.2 of the NEC specifies which locations in service stations are considered classified. Wiring in these areas must be in accordance with their class and division.

Underground wiring must be in either threaded rigid metal conduit or threaded intermediate metal conduit. Any portions of the facility located beneath Class 1, Division 2 areas must be considered Class 1, Division 1 locations and shall be considered as such up to the point at which the raceway emerges from the ground or floor. Properly installed Type MI cable is also permitted. Also, rigid non-metallic conduit can be used if installed at least 2 feet below grade *and* if rigid metal conduit is used for the last 2 feet of the run prior to its emergence from the ground or floor.

Bulk Storage Plants

The requirements for bulk storage facilities are found in *Article 515* of the NEC, as follows:

Installation Methods

Table 515.2 of the NEC specifies which locations in bulk storage plants are considered classified. Wiring in these areas must be in accordance with their class and division.

Underground wiring must be in either threaded rigid metal conduit or threaded intermediate metal conduit. Any portions of the plant located beneath Class 1, Division 2 areas must be considered Class 1, Division 1 locations and shall be considered as such up to the point at which the raceway emerges from the ground or floor. Properly installed Type MI cable is also permitted. Also, rigid non-metallic conduit can be used if installed at least 2 ft below grade *and* if rigid metal conduit is used for the last 2 ft of the run prior to its emergence from the ground or floor (see Figs. 14-15 through 14-20).

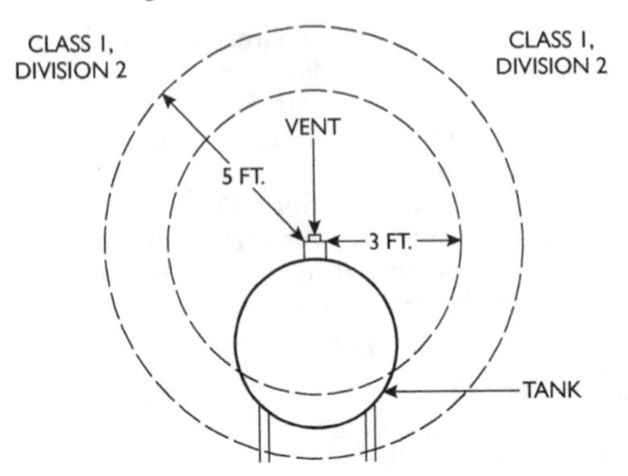

Fig. 14-15 Class 1 areas.

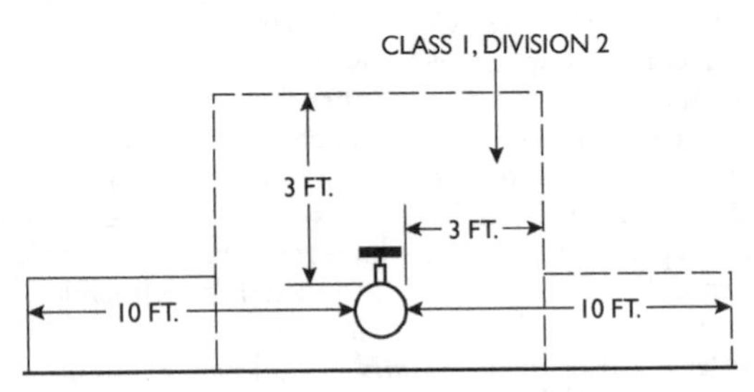

Fig. 14-16 Class 1 areas.

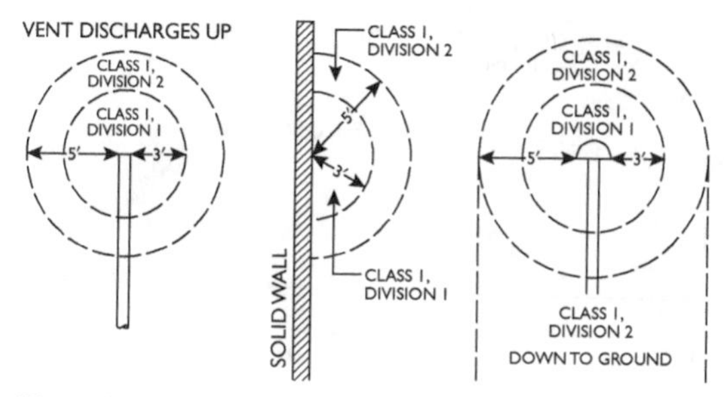

Fig. 14-17 Class I areas around vent pipes.

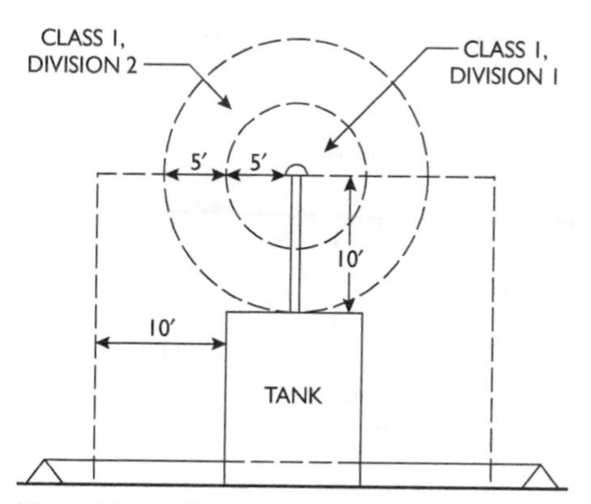

Fig. 14-18 Class I areas around tanks.

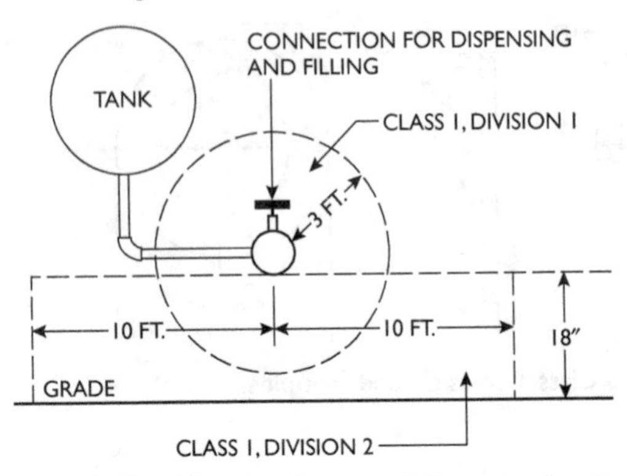

Fig. 14-19 Class I areas around dispensing locations.

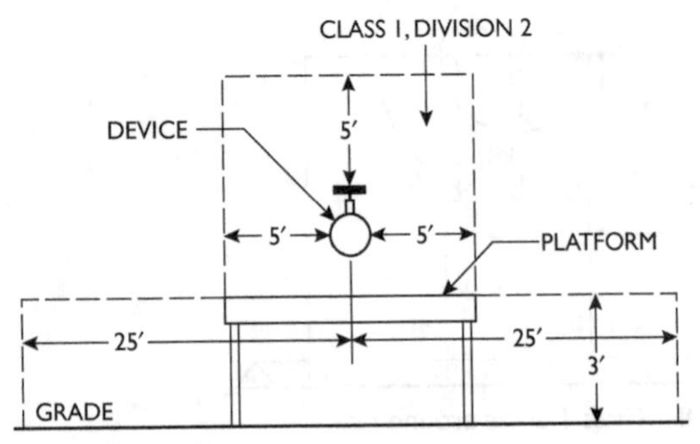

Fig. 14-20 Class I locations.

All wiring above Class 1 locations must be done by one of the following methods:

1. Rigid metal conduit
2. Intermediate metal conduit
3. Electrical metallic tubing
4. Schedule 80 PVC conduit
5. Type MI, T, SNM, or MC cable

Spray Areas

Spray areas are hazardous because of the high concentrations of solvents used for spray processes (see special lighting method in Fig. 14-21).

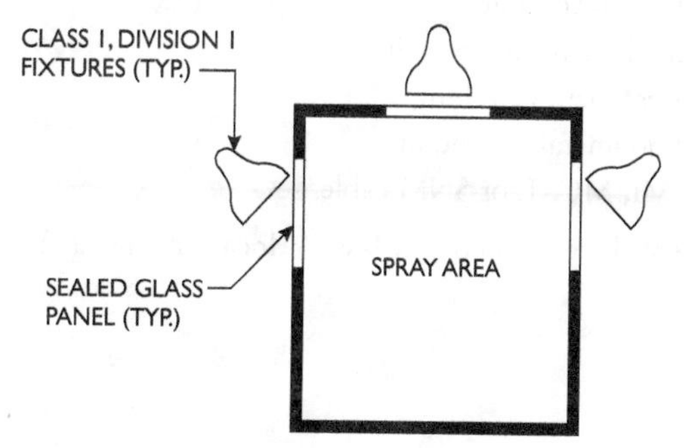

Fig. 14-21 Lighting for spray areas.

The following special requirements for spray areas are found in *Article 516* of the NEC:

Installation Methods

Figures 1, 2, 3, 4, and 5 of *Article 516.3(B)* of the NEC specify which locations in spray areas are considered classified. Wiring in these areas must be in accordance with their class and division.

Adjacent areas such as storerooms and switchboard rooms are not considered to be classified areas if they are well separated from the spray area by tight walls or partitions without communicating openings.

All wiring above Class 1 locations must be done by one of the following methods:

1. Rigid metal conduit
2. Intermediate metal conduit
3. Electrical metallic tubing
4. Rigid nonmetallic conduit
5. Type MI, MC, T, or SNM cable

Equipment located above Class 1 locations must be totally enclosed.

15. TOOLS AND SAFETY

Safety Requirements

Since the late 1980s, there has been a large increase in the number and types of safety regulations placed on electrical installations. (Before that time we had difficulties with OSHA and other organizations, but not on the same scale as we are seeing them now.) These regulations have sprung from our political and legal systems and have caused quite a bit of extra work for all types of building contractors. You will find such regulations and requirements in almost all new contracts, and even on many purchase orders and other less formal commercial agreements. The owners and suppliers have no choice but to write them into their contract documents. If they did not, they would leave themselves open to liability lawsuits and government penalties.

You must pay attention to these requirements; they are being enforced more strictly all the time.

The documentation and handling of hazardous materials (often shortened to "hazmat") are largely a paperwork problem. The amount of paperwork, however, depends on the types of materials that will be used in the project. A typical hazardous material information sheet is shown in Fig. 15-1. (Sometimes, you may be able to obtain completed sheets from the suppliers that sell you the materials.) The hazmat regulations require you to have one person in your company responsible for all of these sheets and records.

Although these regulations seem rather needless, don't underestimate the power of the Environmental Protection Agency (EPA) to enforce them. Like OSHA (but even more), EPA can penalize your company heavily for even a minor infraction.

Section I **MATERIAL SAFETY DATA SHEET**

Supplier's Name: ______ Emergency Phone Number: ______

Address: ______

Chemical Name: LIQUEFIED PETROLEUM GAS or PROPANE Formula: C_3H_8

CAS Registry No.: 74-98-6 Chemical Family: Hydrocarbon

WARNING: Danger! Extremely flammable. Compressed Gas Asphyxiant in high concentrations. Contact with liquid causes burns similar to frost bite. OSHA permissible exposure limit (PEL) 1000 ppm for an 8-hour workday.

Appearance and Odor: Vapor and liquid are colorless. Product contains an odorant (unpleasing odor).

FIRE HAZARD: 4
HEALTH HAZARD: 1
REACTIVITY: 0

4 – Severe
3 – Serious
2 – Moderate
1 – Slight
0 – Minimal

Section II **HAZARDOUS INGREDIENTS**

Hazardous Mixtures: Air with 2.15 to 9.60 percent propane

Section III **PHYSICAL DATA**

Boiling Point: – 44°F
Specific Gravity (H_2O=1): 0.51
Vapor Density (air = 1): 1.52
Solubility in Water: Slightly

Vapor Pressure (PSIG) at 100°F: 205
Percent, Volatile by Volume (%): 100
Evaporation rate: Gas at normal ambient temperatures

Section IV **FIRE AND EXPLOSION HAZARD DATA**

Flash Point: –156°F (CC) Classification: Flammable Gas UN 1075
Flammable Limits - LFL: 2.15 UFL: 9.60 Extinguishing Media: Water spray-Class A-B-C or BC fire extinguisher.
Special Fire Fighting Procedures: Stop flow of gas. Use water to keep fire exposed containers cool. Use water spray to disperse unignited gas or vapor. If ignition has occurred and no water available, tank metal may weaken from overheating. Evacuate area. If gas has not ignited, LP-gas liquid or vapor may be dispersed by water spray or flooding.

Decomposition Products under Fire Conditions: Fumes, smoke, carbon monoxide, aldehydes and other decomposition products, in the case of incomplete combustion or when used as an engine fuel.

"EMPTY" Container Warning: "Empty" containers retain residue (liquid and/or vapor) and can be dangerous. DO NOT PRESSURIZE, CUT, WELD, BRAZE, SOLDER, DRILL, GRIND OR EXPOSE SUCH CONTAINERS TO HEAT, FLAME, SPARKS OR OTHER SOURCES OF IGNITION; THEY MAY EXPLODE AND CAUSE INJURY OR DEATH. Do not attempt to clean since residue is difficult to remove. All containers should be disposed of in an evironmentally safe manner and in accordance with governmental regulations.

Section V **HEALTH HAZARD**

OSHA P.E.L.: 1000 PPM ACGJH TLV: 1000 PPM Effects of Overexposure: Inhalation - concentrations can lead to symptoms ranging from dizziness to anesthesia and respiratory arrest. Eyes - moderate irritation. Emergency & First Aid procedures: Inhalation - remove to fresh air. Guard against self-injury. Apply artificial respiration if breathing has stopped.

(continues)

Fig. 15-1 Hazmat sheet.

Section VI **REACTIVITY DATA**

Stable: X Unstable: ____ Hazardous Decomposition Products: ____ None: ____
Incompatibility (materials to avoid): Mixing with oxygen or air, except at burner
Hazardous Polymerization: May occur ____ Will not occur X

Section VII **SPILL OR LEAK PROCEDURES**

Steps to be taken in case material is released: Keep public away. Shut off supply of gas. Eliminate sources of ignition. Ventilate the area. Disperse with water spray. Contact between skin and these gases in liquid form can cause freezing of tissue causing injury similar to thermal burn.
Waste Disposal Method: Controlled burning. Contact supplier.

Section VIII **SPECIAL PROTECTION INFORMATION**

Respiratory Protection: Stay out of gas or vapor (because of fire hazard). Ventilation: Explosion-proof motors and keep sources of ignition at safe distances. Personal Protective Equipment and Apparel: Gloves resistant to the actions of LP-gases, goggles for protection against accidental release of pressurized products.

Section IX **SPECIAL PRECAUTIONS**

Precautions to be taken when handling and storing: Keep containers away from heat sources and store in upright position. Containers should not be dropped. Keep container valve closed when not in use. Other Precaution: Install protective caps and plug container service valve when not connected for use.

Section X **TOXICOLOGICAL INFORMATION**

OSHA Carcinogen Classification (29 CFR 1910) Not listed/applicable X
U.S. Department of Health (21 CFR 184.1655): Generally recognized as safe (GRAS) as a direct human food ingredient when used as a propellant, aerating agent and gas as defined in Section 170.3(o)(25).

Section XI **DOT LABELING INFORMATION (49 CFR 100-199)**

Proper shipping name: Liquefied Petroleum Gas Identification No.: UN 1075
Hazardous Classification: Flammable Gas Label(s) Required: Flammable Gas

Section XII **ISSUE INFORMATION**

Issue Date: ____

This material safety data sheet and the information it contains is offered to you in good faith as accurate. This Supplier does not manufacture this product but is a supplier of the product independently manufactured by others. Much of the information contained in this data sheet was received from sources outside our Company. To the best of our knowledge this information is accurate, but this Supplier does not guarantee its accuracy or completeness. Health and safety precautions in this data sheet may not be adequate for all individuals and/or situations. It is the user's obligation to evaluate and use this product safely, comply with all applicable laws and regulations and to assume the risks involved in the use of this product.

NO WARRANTY OR MERCHANTABILITY, FITNESS FOR ANY PARTICULAR PURPOSES, OR ANY OTHER WARRANTY IS EXPRESSED OR IS TO BE IMPLIED REGARDING THE ACCURACY OR COMPLETENESS OF THIS INFORMATION, THE RESULTS TO BE OBTAINED FROM THE USE OF THIS INFORMATION OR THE PRODUCT, THE SAFETY OF THIS PRODUCT, OR THE HAZARDS RELATED TO ITS USE.

Fig. 15-1 Hazmat sheet *(continued)*.

OSHA safety regulations also have become very strict in recent years and will probably become more so. Many projects will require you to appoint an employee on the jobsite as safety coordinator. Usually this will require a separate employee on the site at all times, to be concerned only with safety matters—not doing any installation of electrical materials. At times, you may be forced to hire an outside contractor to serve as the safety officer. In either event, you need to be aware of such requirements and to account for the associated expenses.

On many projects (more commonly the larger ones), you will be required to send your workers to safety training. This will require not only training expenses (paid to an outside training company) but also a loss of productivity, since the workers will have to attend several days' worth of classes during normal working hours. They will be required to go to a training center instead of going to work for several days. You may also have to give them training upgrades every so often, and may be subject to other requirements.

This is not a pretty situation from the electrician's point of view, but it is the reality with which we must deal. You must ascertain how these requirements affect the projects you run. Safety training regulations are a very serious business and can give you plenty of grief if you don't make allowances for them beforehand. Even the lack of a few pairs of safety glasses can cost you. And the money gets really deep when your people must work where asbestos or some other hazardous material is present. They won't only have to undergo training, but will also have to wear Nomex suits (which are very expensive) and respirators. Don't expect your workers to get as much done when they are in full-body suits, breathing through something similar to a gas mask.

Tools

Tools are a chief concern for good electrical installations, both for the sake of time saving and also to increase the quality of the installation.

Obviously, the most common tools used in electrical work are the hand tools, most typically the following:

- Hammer (20-oz)
- Lineman's pliers (9-in.)
- "T-type" cable and wire strippers
- Blade screwdriver (large)
- Blade screwdriver (small)
- Phillips screwdriver (large)
- Phillips screwdriver (small)
- Channel-lock pliers (10-in.)
- Hex wrench set
- Razor knife (retractable)
- Hacksaw frame
- Voltage tester ("Wiggy" type)
- Flashlight with continuity tester attachment
- Carpenter's apron
- Tape measure (1-in. × 25-ft)
- Pencils and markers
- Long-nose pliers

Beyond the tools that each electrician is expected to carry, other tools and equipment will be required to perform many installations. The most common types include:

- Right-angle drill
- ⅜-in. variable-speed reversing drill
- Cordless drill and extra battery
- Drill bits:

    1-in. auger bit

    2-in. auger bit

    ¼-in. masonry bit

    Small set of high-speed steel bits

- Circular saw
- Extension cords
- 6-ft step ladder
- 20-ft extension ladder
- Medium-size shovel
- Post-hole digger
- Fish tape
- Hacksaw blades
- Wire lube
- PVC glue
- Electrical tape
- Other power tools

Without question, power tools are important to most electrical installations. Of note in recent years has been the use of cordless tools. When cordless tools were first introduced a number of years ago, they had obvious advantages (e.g., no cord). But they needed frequent charging, took a long time to charge, and were lacking in power. The new models, however, have pretty well conquered these difficulties. Cordless drills and screwdrivers have proven especially useful.

The best method of dealing with cordless tools is to buy an extra battery pack with each cordless tool you buy. By doing so, you will have two batteries for each tool and can have one battery charging while the other battery is in use. Some people also buy extra chargers. It has proven helpful to set up a central charging center for large projects, with one person left in charge of overseeing the supply of charged batteries. This arrangement may not work well for most electrical projects but can work very well on some.

Maintenance

The care that electricians give to their power tools has often been lacking. It is important to keep all your tools in good working condition. An electrician using a drill motor with a defective switch cannot do his best work if he has to operate the switch several times to make contact each time he uses it.

It is particularly important that the electrician treats the tool well, not throw it around or yank at the cord, for example. For large shops or large jobsites, it would be best to standardize on one brand of each type of tool and then to keep spare parts on hand for each model.

Special Power Tools

The field of portable power tools covers everything from small electric hand tools to heavy-duty drilling, grinding, and driving tools. Most mechanics have experience with the more common types and are knowledgeable in their correct usage and safe operation. There are, however, some more specialized types that mechanics may use only occasionally. Because these specialized tools, as well as all other power tools, operate at relatively high speed and may use sharp-edged cutters, safe and efficient operation requires knowledge and understanding of both the power unit and the auxiliary parts.

Hammer Drills

The hammer drill combines conventional rotary drilling with hammer or percussion drilling. Its primary use is to drill holes in concrete or masonry. It incorporates two-way action, in that it may be set for hammer drilling with rotation or for rotation only. When drilling in concrete or masonry, special carbide-tipped bits are required. These are made with alloy-steel shanks (for durability) onto which carbide tips are brazed to provide the hard cutting edges necessary to resist dulling. The concrete or masonry is reduced to granules and dust by the combined hammering and rotating action, and shallow spiral flutes on the drill bit remove material from the hole.

Core Drills

Prior to the introduction of concrete core drilling, holes in concrete structures required either careful planning and preparation or breaking away sections of hard concrete and considerable patching. The tool that has changed these practices is the core bit. The core bit is basically a metal tube with a crown imbedded with industrial-type diamonds or similar abrasive substances on one end. Another type of diamond bit has all the diamonds set on the surface of the bit crown and arranged in a predetermined pattern for maximum cutting effectiveness. Bits are made in two styles:, the closed-back style and the open-back. These are shown in Fig. 15-2.

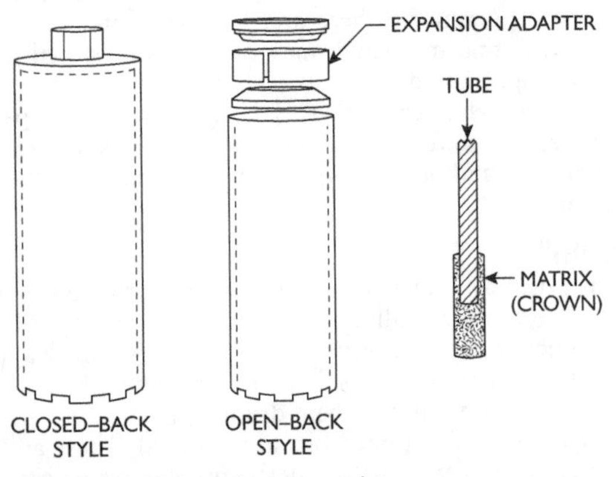

Fig. 15-2 Diamond-type core bits.

There is a cost advantage in the use of the open-back bit, in that the reusable adapters offer savings in cost on each bit after the first. Another, sometimes appreciable advantage is that should a core become lodged in the bit, removing the adapter makes core removal easier. The closed-back bit offers the advantage of simplicity. Installation requires only turning the bit onto the arbor thread, with no problem of positioning or alignment. Because the bit is a single, complete unit, there is no problem with mislaid, lost, or damaged parts.

Successful core drilling requires that several very important conditions be maintained: rigidity of the drilling unit, adequate water flow, and uniform steady pressure. These are illustrated in Fig. 15-3. Understanding the action that takes place when a diamond core bit is in operation will result in better appreciation of the importance of maintaining these conditions.

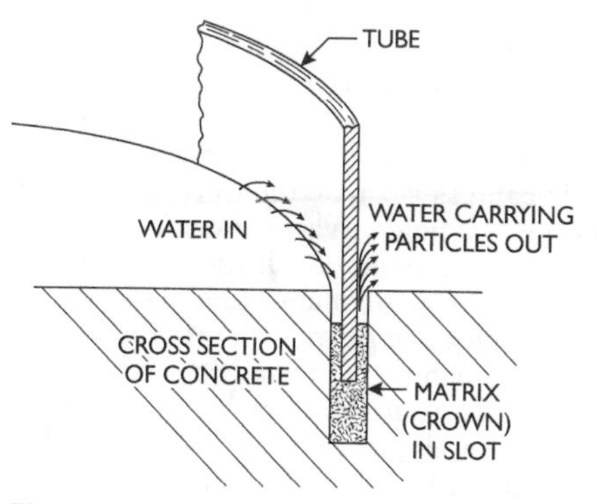

Fig. 15-3 Core drilling.

The core drilling machine is, in effect, a special drill press. The power unit is mounted in a cradle that is moved up and down the column by moving the operating handles. The handles rotate a pinion gear that meshes with a rack attached to the column. To secure the rig to the work surface, the top of the column is provided with a jack screw that locks the top of the column against the overhead with the aid of an extension. Some concrete core drilling rigs are equipped with vacuum systems. This makes it possible to attach the rig directly to the work surface.

The power unit is a heavy-duty electric motor with reduction gears to provide steady rotation at the desired speed. The motor spindle incorporates a water swivel, which allows introduction of water through a hole in the bit adapter to the inside of the core bit.

Portable Band Saws

The portable band saw allows the electrician to perform many cutting operations efficiently and with little effort. Many field operations that require cutting in place consume excessive time and effort if done by hand. If acetylene torch cutting were used, the resulting rough surfaces would not be acceptable, and/or the flame and flying molten metal might be hazardous. There are many times when the object being worked on (almost always a medium-to-large conduit) cannot be taken to another location to be sawed. In such instances, a portable band saw can do the work in the field. The portable band saw is a lightweight, self-contained small version of the standard shop band saw. It uses two rotating rubber-tired wheels to drive a continuous saw blade. Power is supplied by a heavy-duty electric motor that transmits its power to the wheels through a gear train and worm wheel reduction. Speed reduction is accomplished in this manner, and powerful torque is developed to drive the saw band at proper speed. Blade guides, similar in design to those on the standard shop machine, are built into the unit, along with a bearing behind the blade to handle thrust loads.

Blades for portable band saws are available in a tooth pitch range from 6 to 24 teeth per inch. The rule of thumb for blade selection is to have three (3) teeth in the material at all times. Using too coarse a blade will cause thin metals to hang up in the gullet between two teeth and tear out a section of teeth. Too fine a blade will prolong the cutting job, because only small amounts of metal will be removed by each tooth. Don't use cutting oil with a band saw. If you do, oil and metal chips will usually build up on the rubber tires, causing the blade to slip under load and causing misalignment of the blade.

Powder-Actuated Tools

The principle of the powder-actuated tool is to fire a fastener into material to anchor or make it secure to another material. Some applications are wood to concrete, steel to concrete, wood to steel, steel to steel, and numerous applications that require fastening fixtures and special articles to concrete or steel. Because the tools vary in design details and safe handling techniques, general information and descriptions of the basic tools and accessories are given, rather than specific operating instructions. Because the principle of operation is similar to that of a firearm (most of them use .22 caliber cartridges), safe handling and use must be given the highest priority. A powder-actuated tool is, in effect, a pistol that fires a round composed of two elements:

1. Cartridge with firing cap and powder
2. Bullet

When you pull the trigger on a pistol, the firing pin detonates the firing cap and powder and sends a bullet in free flight to its destination. The direct-acting type of powder-actuated tool can be described in the same manner, with one small change. The bullet is loaded as two separate parts. Fig. 15-4 shows the essential parts of a powder-actuated tool system—the tool and the two-part bullet (the combination of the fastener and the powder load).

In operation, the fastener and powder load are inserted in the tool, the tool is pressed against the work surface, the trigger is pulled, and the fastener travels in free flight to its destination.

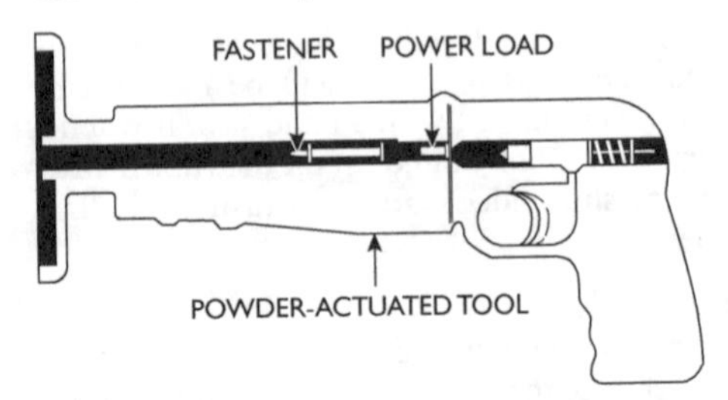

Fig. 15-4 Parts of a powder-actuated tool.

Although different manufacturers' tools may vary in appearance, they are all similar in principle and basic design. They all have, within a suitable housing, a chamber to hold the powder load, a firing mechanism, a barrel to confine and direct the fastener, and a shield to confine flying particles. There are two types of tools:

Direct-acting (shown in Fig. 15-5), in which the expanding gas acts directly on the fastener to be driven into the work.

Indirect-acting (shown in Fig. 15-6), in which the expanding gas of a powder charge acts on a captive piston that drives the fastener into the work.

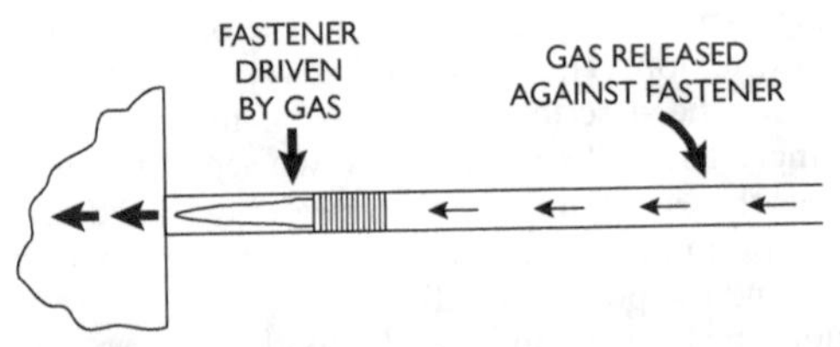

Fig. 15-5 Direct-acting fastening.

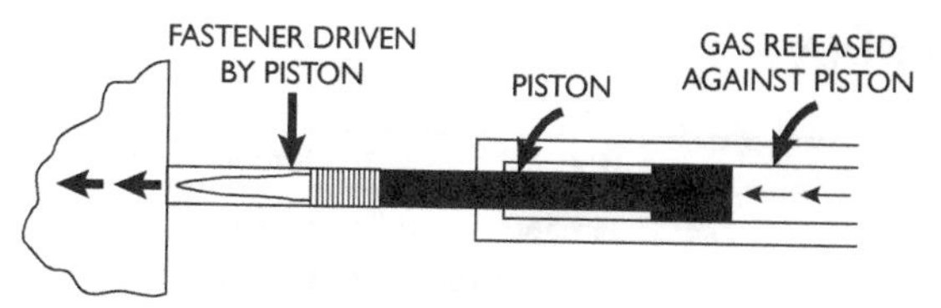

Fig. 15-6 Indirect-acting fastening.

A variety of fasteners are used with powder-actuated tools. Typically in the form of a large pin, they are made of special grades of steel that makes it possible for them to penetrate concrete or steel without breaking. Some of these fasteners have a head on them, like a nail; some are tipped with threaded studs; and other types can be used also.

It is important always to match the fastener and the material it is being fastened to with an appropriate powder load. In other words, you need the right amount of gunpowder for the purpose. Too small a powder load won't fasten the work correctly, and too large a powder load can be dangerous. All manufacturers' instructions with fasteners identify the proper powder load for their products. Typically, powder loads are rated 1 to 12, with 12 being the most powerful.

When using powder-actuated fasteners with concrete, it is important not to drive the fastener too close to an edge of the concrete. Doing so may very well blow out the edge of the concrete.

APPENDIX

Multiply	*By*	*To Obtain*
Acres	43,560	Square feet
Acres	1.562×10^{-3}	Square miles
Acre-feet	43,560	Cubic feet
Amperes per sq cm	6.452	Amperes per sq in
Amperes per sq in.	0.1550	Amperes per sq cm
Ampere-turns	1.257	Gilberts
Ampere-turns per cm	2.540	Ampere-turns per in.
Ampere-turns per in.	0.3937	Ampere-turns per cm
Atmospheres	76.0	Cm of mercury
Atmospheres	29.92	Inches of mercury
Atmospheres	33.90	Feet of water
Atmospheres	14.70	Pounds per sq in.
British thermal units (Btu)	252.0	Calories
British thermal units	778.2	Foot-pounds
British thermal units	3.930×10^{-4}	Horsepower-hours
British thermal units	0.2520	Kilogram-calories
British thermal units	107.6	Kilogram-meters
British thermal units	2.931×10^{-4}	Kilowatt-hours
British thermal units	1,055	Watt-seconds
Btu per hour	2.931×10^{-4}	Kilowatts
Btu per minute	2.359×10^{-2}	Horsepower
Btu per minute	1.759×10^{-2}	Kilowatts
Bushels	1.244	Cubic feet
Centimeters	0.3937	Inches
Circular mils	5.067×10^{-6}	Square centimeters
Circular mils	0.7854×10^{-6}	Square inches

Multiply	*By*	*To Obtain*
Circular mils	0.7854	Square mils
Cords	128	Cubic feet
Cubic centimeters	6.102×10^{-2}	Cubic inches
Cubic feet	0.02832	Cubic meters
Cubic feet	7.481	Gallons
Cubic feet	28.32	Liters
Cubic inches	16.39	Cubic centimeters
Cubic meters	35.31	Cubic feet
Cubic meters	1.308	Cubic yards
Cubic yards	0.7646	Cubic meters
Degrees (angle)	0.01745	Radians
Dynes	2.248×10^{-6}	Pounds
Ergs	1	Dyne-centimeters
Ergs	7.37×10^{-6}	Foot-pounds
Ergs	10^{-7}	Joules
Farads	10^{6}	Microfarads
Fathoms	6	Feet
Feet	30.48	Centimeters
Feet of water	0.8826	Inches of mercury
Feet of water	304.8	Kg per square meter
Feet of water	62.43	Pounds per sq foot
Feet of water	0.4335	Pounds per sq in.
Foot-pounds	1.285×10^{-2}	British thermal units
Foot-pounds	5.050×10^{-7}	Horsepower-hours
Foot-pounds	1.356	Joules
Foot-pounds	0.1383	Kilogram-meters
Foot-pounds	3.766×10^{-7}	Kilowatt-hours
Gallons	0.1337	Cubic feet

(continued)

Multiply	*By*	*To Obtain*
Gallons	231	Cubic inches
Gallons	3.785×10^{-3}	Cubic meters
Gallons	3.785	Liters
Gallons per minute	2.228×10^{-3}	Cubic feet per sec
Gausses	6.452	Lines per square in.
Gilberts	0.7958	Ampere-turns
Henries	10^3	Millihenries
Horsepower	42.41	Btu per minute
Horsepower	2,544	Btu per hour
Horsepower	550	Foot-pounds per sec
Horsepower	33,000	Foot-pounds per min
Horsepower	1.014	Horsepower (metric)
Horsepower	10.70	Kg-calories per min
Horsepower	0.7457	Kilowatts
Horsepower (boiler)	33,520	Btu per hour
Horsepower-hours	2,544	British thermal units
Horsepower-hours	1.98×10^6	Foot-pounds
Horsepower-hours	2.737×10^5	Kilogram-meters
Horsepower-hours	0.7457	Kilowatt-hours
Inches	2.54	Centimeters
Inches of mercury	1.133	Feet of water
Inches of mercury	70.73	Pounds per square ft
Inches of mercury	0.4912	Pounds per square in.
Inches of water	25.40	Kg per square meter
Inches of water	0.5781	Ounces per square in.
Inches of water	5.204	Pounds per square ft
Joules	9.478×10^{-4}	British thermal units
Joules	0.2388	Calories
Joules	10^7	Ergs

Multiply	*By*	*To Obtain*
Joules	0.7376	Foot-pounds
Joules	2.778×10^{-7}	Kilowatt-hours
Joules	0.1020	Kilogram-meters
Joules	1	Watt-seconds
Kilograms	2.205	Pounds
Kilogram-calories	3.968	British thermal units
Kilogram-meters	7.233	Foot-pounds
Kg per square meter	3.281×10^{-3}	Feet of water
Kg per square meter	0.2048	Pounds per sq ft
Kg per square meter	1.422×10^{-3}	Pounds per sq in.
Kilolines	10^3	Maxwells
Kilometers	3.281	Feet
Kilometers	0.6214	Miles
Kilowatts	56.87	Btu per min
Kilowatts	737.6	Foot-pounds per sec
Kilowatts	1.341	Horsepower
Kilowatt-hours	3.412	British thermal units
Kilowatt-hours	2.655×10^6	Foot-pounds
Knots	1.152	Miles
Liters	0.03531	Cubic feet
Liters	61.02	Cubic inches
Liters	0.2642	Gallons
Log_e N	0.4343	Log_{10}N
Log N	2.303	Log_e N
Lumens per square ft	1	Foot-candles
Maxwells	10^{-3}	Kilolines
Megalines	10^6	Maxwells
Megohms	10^6	Ohms

(continued)

Multiply	*By*	*To Obtain*
Meters	3.281	Feet
Meters	39.37	Inches
Meter-kilograms	7.233	Pound-feet
Microfarads	10^{-6}	Farads
Microhms	10^{-6}	Ohms
Microhms per cm cube	0.3937	Microhms per in. cube
Microhms per cm cube	6.015	Ohms per mil foot
Miles	5,280	Feet
Miles	1.609	Kilometers
Miner's inches	1.5	Cubic feet per min
Ohms	10^{-6}	Megohms
Ohms	10^{6}	Microhms
Ohms per mil foot	0.1662	Microhms per cm cube
Ohms per mil foot	0.06524	Microhms per in. cube
Pounds	0.03108	Pounds
Pounds	32.17	Poundals
Pound-feet	0.1383	Meter-kilograms
Pounds of water	0.01602	Cubic feet
Pounds of water	0.1198	Gallons
Pounds per cubic foot	16.02	Kg per cubic meter
Pounds per cubic foot	5.787×10^{-4}	Pounds per cubic in.
Pounds per cubic inch	27.68	Grams per cubic cm
Pounds per cubic inch	2.768×10^{-4}	Kg per cubic meter
Pounds per cubic inch	1.728	Pounds per cubic ft
Pounds per square foot	0.01602	Feet of water
Pounds per square foot	4.882	Kg per cubic meter
Pounds per square foot	6.944×10^{-3}	Pounds per sq in.
Pounds per square inch	2.307	Feet of water

Multiply	*By*	*To Obtain*
Pounds per square inch	2.036	Inches of mercury
Pounds per square inch	703.1	Kg per square meter
Radians	57.30	Degrees
Square centimeters	1.973×10^{-5}	Circular mils
Square feet	2.296×10^{-5}	Acres
Square feet	0.09290	Square meters
Square inches	1.273×10^{6}	Circular mils
Square inches	6.452	Square centimeters
Square kilometers	0.3861	Square miles
Square meters	10.76	Square feet
Square miles	640	Acres
Square miles	2.590	Square kilometers
Square millimeters	1.937×10^{3}	Circular mils
Square mils	1.273	Circular mils
Tons (long)	2,240	Pounds
Tons (metric)	2,205	Pounds
Tons (short)	2,000	Pounds
Watts	0.05686	Btu per minute
Watts	10^{7}	Ergs per sec
Watts	44.26	Foot-pounds per min
Watts	1.341×10^{-3}	Horsepower
Watts	14.34	Calories per min
Watt-hours	3.412	British thermal units
Watt-hours	2,655	Foot-pounds
Watt-hours	1.341×10^{-3}	Horsepower-hours
Watt-hours	0.8605	Kilogram-calories
Watt-hours	376.1	Kilogram-meters
Webers	10^{8}	Maxwells

Index

B

C

F

R

S

U

Z